OCR GCSE MATHEMATICS

STAGES

9
10

SI ... AGER

GRADUATED ASSESSMENT

SECOND EDITION

- Howard Baxter
- Michael Handbury
- John Jeskins
- Jean Matthews
- Mark Patmore

Hodder Education

A MEMBER OF THE HODDER HEADLINE GROUP

The Publishers would like to thank the following for permission to reproduce copyright material:

Photo credits: p. 141 © George McCarthy/Corbis; p.178 © Robert Gill; Papilio/Corbis.

Acknowledgements Every effort has been made to trace all copyright holders, but if any have been inadvertently overlooked the Publishers will be pleased to make the necessary arrangements at the first opportunity.

Hodder Headline's policy is to use papers that are natural, renewable and recyclable products and made from wood grown in sustainable forests. The logging and manufacturing processes are expected to conform to the environmental regulations of the country of origin.

Orders: please contact Bookpoint Ltd, 130 Milton Park, Abingdon, Oxon OX14 4SB. Telephone: (44) 01235 827720. Fax: (44) 01235 400454. Lines are open from 9 a.m. to 5 p.m., Monday to Saturday, with a 24-hour message-answering service. Visit our website at www.hoddereducation.co.uk.

© Howard Baxter, Michael Handbury, John Jeskins, Jean Matthews, Mark Patmore, Brian Seager, Eddie Wilde, 2007
First published in 2007 by
Hodder Education,
part of Hachette Livre UK,
338 Euston Road,
London, NW1 3BH

Impression number   10 9 8 7 6 5 4 3
Year   2012 2011 2010 2009 2008

Cover photo © Andy Sacks/Photographer's Choice/Getty Images
Illustrations © Barking Dog Art
Typeset in Futura Book 12/14pt by Pantek Arts Ltd, Maidstone, Kent
Printed and bound in Malaysia for Hodder Education, a division of Hodder Headline plc, 338 Euston Road, London NW1 3BH.

A catalogue record for this title is available from the British Library.

ISBN: 978 0340 915 950

Stage 9 Contents

STAGE
9

STAGE
9

Stage 9 Contents

iv

Introduction

About this book

This course has been written especially for students following OCR's 2006 Modular Specification C, Graduated Assessment (J516) for GCSE Mathematics.

This book covers the complete specification for Stages 9 and 10.

- Each chapter is presented in a way which will help you to understand the mathematics, with straightforward explanations and worked examples covering every type of problem.
- At the start of each chapter are two lists, one of what you should already know before you begin and the other of the topics you will be learning about in that chapter.
- 'Activities' offer a more interesting approach to the core content, giving opportunities for you to develop your skills.
- 'Challenges' are rather more searching and are designed to make you think mathematically.
- There are plenty of exercises to work through to practise your skills.
- Some questions are designed to be done without a calculator, so that you can practise for the non-calculator sections of the examination papers.
- Look out for the 'Exam tips' – these give advice on how to improve your performance in the module tests, direct from the experienced examiners who have written this book.
- At the end of each chapter there is a short summary of what you have learned.
- Finally, there are 'Revision exercises' at intervals throughout the book to help you revise all the topics covered in the preceding chapters.

Other components in the series

- A Homework Book
  This contains parallel exercises to those in this book to give you more practice. Included with the Homework Book is a Personal Tutor CD-ROM. This will help you if you have to miss a lesson or if you need a reminder of something taught in class.

- An Assessment Pack

  There are two Assessment Packs: one for Foundation Tier (Stages 1 to 7) and one for Higher Tier (Stages 6 to 10). Each contains revision exercises, practice module papers and a practice terminal paper to help you prepare for the examination. Some of the questions in the examination will offer you little help to get started. These are called 'unstructured' or 'multi-step' questions. Instead of the question having several parts, each of which helps you to answer the next, you have to work out the necessary steps to find the answer. There are examples of this kind of question in the Assessment Pack.

- An Interactive Investigations CD-ROM

  This contains whole-class presentations and individual activities. It helps you understand how you can best use ICT to do your homework and other tasks.

Top ten tips

Here are some general tips from the examiners who wrote this book to help you to do well in your tests and examinations.

Practise

1 **taking time** to work through each question carefully.
2 answering questions **without** a calculator.
3 answering questions which require **explanations**.
4 answering **unstructured** questions.
5 **accurate** drawing and construction.
6 answering questions which **need a calculator**, trying to use it efficiently.
7 **checking answers**, especially for reasonable size and degree of accuracy.
8 making your work **concise** and well laid out.
9 checking that you have **answered the question**.
10 **rounding** numbers, but only at the appropriate stage.

Checking answers

You will learn about

- Checking answers by rounding to 1 significant figure
- Simplifying calculations using standard form

You should already know

- How to interpret significant figures
- How to write numbers in standard form

Checking answers by rounding to 1 significant figure

It is important to be able to check calculations quickly, without using a calculator. One way to do this is to round the numbers to 1 significant figure.

EXAMPLE 1

Find an approximate answer to this calculation.
5.13×4.83

Round 5.13 and 4.83 each to 1 significant figure to give a much simpler calculation

$$5.13 \times 4.83 \approx 5 \times 5$$
$$= 25$$

EXAM TIP

In a calculation it may be possible to round one number up and another number down. This is more likely to give an answer close to the exact answer than rounding both numbers up or down.

STAGE
9

EXAMPLE 2

Find an approximate answer to this calculation.

$$\frac{(3\cdot26 \times 10^3) \times (8\cdot17 \times 10^5)}{6\cdot28 \times 10^2}$$

$$\frac{(3\cdot26 \times 10^3) \times (8\cdot17 \times 10^5)}{6\cdot28 \times 10^2} \approx \frac{3 \times 8}{6} \times \frac{10^3 \times 10^5}{10^2}$$

Round numbers to 1 significant figure.
Collect together numbers and powers of 10.

$$= \frac{24}{6} \times \frac{10^8}{10^2}$$

Add indices when multiplying powers of 10.

$$= 4 \times 10^6$$

Subtract indices when dividing powers of 10.

ACTIVITY 1

Without working them out, write down whether or not each of these calculations is correct.

Give your reason in each case.

a) $1975 \times 43 = 84920$

b) $697 \times 0\cdot72 = 5018\cdot4$

c) $3864 \div 84 = 4\cdot6$

d) $19 \times 37 = 705$

e) $306 \div 0\cdot6 = 51$

f) $6127 \times 893 = 54714\cdot11$

Compare your reasons with the rest of the class.

Did you all give the same reasons?

Did anyone have ideas which you hadn't thought of and which you think work well?

EXERCISE 1.1

1 Find an approximate answer to each of these calculations by rounding each number to 1 significant figure.

a) $498 \times 2·18$ b) $13·92 \div 4·8$ c) $4·19 \times 6·68$

d) $881 \div 99$ e) $7·2 \times 9·7$ f) $105·6 \div 5·12$

g) $313 \times 0·68$ h) $4·189 \div 0·477$

Now use a calculator to see how close your approximations are to the correct answers.

2 Find an approximate answer to each of these calculations by rounding each number to 1 significant figure.

a) $159·65 \div 515$ b) $36·8 \times (5·7 + 6·4)$ c) $\sqrt{41\,300}$

d) $0·143 \div 0·116$ e) $(5·67 - 3·85) \times 39$ f) $(34·2)^2$

g) 972×18 h) $0·39^2$ i) $^-19·6 \div 5·2$

Now use a calculator to see how close your approximations are to the correct answers.

3 Find an approximate answer to each of these calculations by rounding each number to 1 significant figure.

a) $\dfrac{2·5 \times 3·6}{5·9}$ b) $\dfrac{0·21 \times 93}{103·1 \div 9·6}$ c) $3·8 \times \sqrt{385}$

d) $\dfrac{543}{18·1} + \dfrac{472}{10·9}$ e) $\dfrac{28·2 \times 3·14}{8·99}$ f) $96·7 \times 4·9^2$

g) $\dfrac{54·3 + 47·2}{9·8 + 10·9}$ h) $\dfrac{\sqrt{5·21 \times 8·35 \times 0·105}}{1·72^2}$

Now use a calculator to see how close your approximations are to the correct answers.

4 Find an approximate answer to each of these calculations by rounding each number to 1 significant figure.

a) $(1·98 \times 10^5) \times (4·65 \times 10^4)$ b) $(1·5 \times 10^8) \times (7·2 \times 10^{-4})$

c) $\dfrac{7·89 \times 10^5}{4·73 \times 10^3}$ d) $(5·59 \times 10^2) \div (1·87 \times 10^5)$

e) $\dfrac{5·84 \times 10^4}{2·68 \times 10^{-2}}$ f) $\dfrac{(8·27 \times 10^{13}) \times (9·75 \times 10^2)}{1·25 \times 10^8}$

g) $\dfrac{(6·89 \times 10^5) \times (7·36 \times 10^{-4})}{4·57 \times 10^{-3}}$ h) $\dfrac{(1·25 \times 10^5)^2}{3·6 \times 10^4}$

i) $(3·2 \times 10^6) \times (9·45 \times 10^4)$ j) $(3·64 \times 10^7) \times (2·4 \times 10^{-5})$

k) $\dfrac{5·93 \times 10^5}{3·29 \times 10^4}$ l) $(3·52 \times 10^4) \div (1·44 \times 10^8)$

m) $\dfrac{8·17 \times 10^{-3}}{1·52 \times 10^{-2}}$ n) $\dfrac{(4·29 \times 10^{-3}) \times (8·18 \times 10^{-5})}{1·5 \times 10^3}$

o) $\dfrac{3·75 \times 10^9}{(5·01 \times 10^{-3}) \times (1·62 \times 10^6)}$ p) $\dfrac{(4·17 \times 10^{-3}) \times (9·29 \times 10^{-5})}{\sqrt{8·63 \times 10^8}}$

Now use a calculator to see how close your approximations are to the correct answers.

EXERCISE 1.1 continued

5 Estimate the answer to each of these calculations.
Show your working.
- **a)** The cost of 7 CDs at £8·99 each.
- **b)** The cost of 29 theatre tickets at £14·50.
- **c)** The cost of 3 meals at £5·99 and 3 drinks at £1·95.

6 a) Use rounding to 1 significant figure to estimate the answer to each of these calculations.
Show your working.

(i) $39·2^3$

(ii) $18·4 \times 0·19$

(iii) $\sqrt{7·1^2 - 3·9^2}$

(iv) $\dfrac{11·6 + 30·2}{0·081}$

b) Use your calculator to find the correct answer to each of the calculations.
Where appropriate, round your answer to a sensible degree of accuracy.

C CHALLENGE 1

In question **6** of Exercise 1.1, the estimates and actual answers agree to 1 significant figure, except for part **(ii)**.

Make up some more estimation and calculation questions of your own.

Try to find other examples where the estimates and accurate answers do not agree to 1 significant figure.

K KEY IDEAS

- Check answers to calculations by rounding numbers to 1 significant figure.

- Use the laws of indices to simplify calculations involving numbers in standard form.

- To multiply numbers in standard form, add the indices.

- To divide numbers in standard form, subtract the indices.

Algebraic manipulation

You will learn about

- Expanding brackets such as $(2x + 3y)(3x + 2y)$
- Simplifying algebraic fractions such as $\dfrac{3x^2y^4}{6xy^2}$ and $\dfrac{x^2 + x}{x^2 + 3x + 2}$
- Factorising expressions such as $20a^2b^3 + 5ab$ and $8x^2 + 4x - 12$
- Recognising and factorising expressions such as $x^2 - y^2$ and $4a^2 - 9b^2$
- Solving quadratic equations of the form $ax^2 + bx + c = 0$ by factorisation

You should already know

- How to expand brackets and manipulate simple algebraic expressions

Multiplying out two brackets

Expressions such as $a(3a - 2b)$ can be multiplied out to give

$a(3a - 2b) = (a \times 3a) - (a \times 2b) = 3a^2 - 2ab.$

This can be extended to working out expressions such as $(2a + b)(3a + b)$.

Each term of the first bracket must be multiplied by each term of the second bracket.

$(2a + b)(3a + b)$

$\quad = 2a(3a + b) + b(3a + b)$ Expand the first bracket.

$\quad = 6a^2 + 2ab + 3ab + b^2$ Notice that the middle two terms are **like terms** and so can be collected.

$\quad = 6a^2 + 5ab + b^2$

In Stage 8 you saw how to multiply out two brackets using a table or the word FOIL. These methods still work for more complicated expressions such as the ones you meet here.

EXAM TIP

As well as being told to multiply out brackets, you may sometimes be asked to simplify, expand or remove the brackets, which all mean the same thing.

STAGE
9

EXAMPLE 1

Multiply out the brackets.

a) $(2a + 3)(a - 1)$

b) $(5a - 2b)(3a - b)$

c) $(2a - b)(a + 2b)$

a) $(2a + 3)(a - 1) = 2a(a - 1) + 3(a - 1)$
$= 2a^2 - 2a + 3a - 3$
$= 2a^2 + a - 3$

Be careful with the signs.

b) $(5a - 2b)(3a - b) = 5a(3a - b) - 2b(3a - b)$
$= 15a^2 - 5ab - 6ab + 2b^2$
$= 15a^2 - 11ab + 2b^2$

Note that it is ^-2b times the bracket.

c) $(2a - b)(a + 2b) = 2a(a + 2b) - b(a + 2b)$
$= 2a^2 + 4ab - ab - 2b^2$
$= 2a^2 + 3ab - 2b^2$

Note that it is ^-b times the bracket.

EXAM TIP
Take care with negative signs. Most errors are made in multiplying out the second bracket when the sign in front is negative.

In each part of Example 1, the two brackets have resulted in three terms.

There are two other types of expansion of two brackets that you need to know about.

EXAMPLE 2

Expand the brackets.

a) $(2a - 3b)^2$

b) $(2a - b)(2a + b)$

EXAM TIP
When squaring a bracket, as in part **a)** of Example 2, make sure that you write the brackets separately and end up with three terms.

a) $(2a - 3b)^2 = (2a - 3b)(2a - 3b)$
$= 2a(2a - 3b) - 3b(2a - 3b)$
$= 4a^2 - 6ab - 6ab + 9b^2$
$= 4a^2 - 12ab + 9b^2$

Note that $^-6ab - 6ab = ^-12ab$

b) $(2a - b)(2a + b) = 2a(2a + b) - b(2a + b)$
$= 4a^2 + 2ab - 2ab - b^2$
$= 4a^2 - b^2$

EXAM TIP
Some people can multiply out two brackets without writing anything down. However, this can lead to errors caused by missing steps and so it is worth showing every step in an examination.

Notice that in part **b)** of Example 2 we get only two terms because the middle terms cancel each other out.

This type of expansion is known as the **difference of two squares** because $(a - b)(a + b) = a^2 - b^2$.

EXERCISE 2.1

Multiply out the brackets.

1 $(x + 2)(x - 3)$

2 $(x + 2)(x + 1)$

3 $(x + 5)(x + 9)$

4 $(x - 3)(x + 6)$

5 $(x - 1)^2$

6 $(x - 4)^2$

7 $(5 + x)(x - 6)$

8 $(2 + x)(5 + x)$

9 $(x - 7)(x + 7)$

10 $(x - 8)(x + 8)$

11 $(5x - 1)(2x - 4)$

12 $(2x - 5)(3x - 2)$

13 $(4x + 2)(3x - 7)$

14 $(5x + 6)(2x - 3)$

15 $(x + y)(2x + y)$

16 $(3x + y)(4x + y)$

17 $(3x - 5y)(x - 4y)$

18 $(2x - 3y)(x - 2y)$

19 $(3x + 4y)(4x - 5y)$

20 $(7x + 8y)(6x - 4y)$

21 $(2g - 3h)(2g - 7h)$

22 $(2h - 7k)(2h - 7k)$

23 $(3j - 8m)(2j - 7m)$

24 $(2k + 7n)(5k - 6n)$

25 $(3p + 8m)(2p - 9m)$

26 $(2r + 3n)(3r - 5n)$

27 $(2q + 7p)(2q - 9p)$

28 $(3r - 8s)(2r - 7s)$

29 $(2s - 3t)(2s - 7t)$

30 $(t - 5)(4t - 3)$

C   CHALLENGE 1

The diagram shows a rectangular garden with a shed in one corner.

(a) Write an expression for
  (i) the length of the garden.
  (ii) the width of the garden.

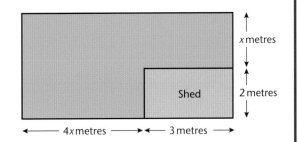

x metres

2 metres

Shed

4x metres   3 metres

(b) Write an expression for the area of the garden.
   Simplify the expression.

C CHALLENGE 2

The diagram shows a rectangular picture hanging on a rectangular wall in a gallery.

The picture is placed so that it is 1 metre from the top of the wall and 1 metre from the bottom of the wall.

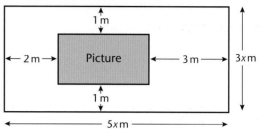

The picture is also 2 metres from the left-hand end of the wall and 3 metres from the right-hand end of the wall.

a) Write an expression for
   (i) the length of the picture.   **(ii)** the width of the picture.

b) Write an expression for the area of the wall occupied by the picture. Simplify the expression.

Simplifying expressions using indices

Remember that $\qquad$ $a \times a \times a = a^3$ and $a \times a \times a \times a \times a = a^5$.

This can be extended $\qquad$ $a^3 \times a^5 = (a \times a \times a) \times (a \times a \times a \times a \times a) = a^8$

which is the same as $\qquad$ $a^3 \times a^5 = a^{3+5} = a^8$.

This suggests a general law for indices.

$$a^m \times a^n = a^{m+n}$$

Similarly, $a^5 \div a^3 = (a \times a \times a \times a \times a) \div (a \times a \times a)$

$$= \frac{a \times a \times a \times a \times a}{a \times a \times a} = a \times a = a^2. \qquad \text{Cancel } a \times a \times a \text{ top and bottom.}$$

This is the same as $a^5 \div a^3 = a^{5-3} = a^2$

which suggests another general law for indices.

$$a^m \div a^n = a^{m-n}$$

Using the first law, you can see that $(a^2)^3 = a^2 \times a^2 \times a^2 = a^6$.

This is the same as $(a^2)^3 = a^{2\times3} = a^6$

and suggests yet another law.

$$(a^n)^m = a^{n \times m}$$

$a^3 \div a^3 = a^{3-3} = a^0$.

But $a^3 \div a^3 = 1$.

This gives another law.

$a^0 = 1$

You can use these laws, together with the algebra you have already learned, to simplify a number of different algebraic expressions.

EXAMPLE 3

Simplify these.

a) $3a^2 \times 4a^3$

b) $\dfrac{6a^5}{2a^3}$

c) $(a^3)^4 \times a^3 \div a^5$

a) $3a^2 \times 4a^3 = 12a^5$

The numbers are multiplied and the indices are added.

b) $\dfrac{6a^5}{2a^3} = 3a^2$

The numbers are divided and the indices are subtracted.

c) $(a^3)^4 \times a^3 \div a^5 = a^{12} \times a^3 \div a^5$
$= a^{15} \div a^5$
$= a^{10}$

Simplify the first term by multiplying the indices.
To multiply, add the indices.
To divide, subtract the indices.

EXAMPLE 4

Simplify where possible.

a) $4a^2b^3 \times 3ab^2$

b) $\dfrac{12ab^3 \times 3a^2b}{2a^3b^2}$

c) $4a^2 + 3a^3$

a) $4a^2b^3 \times 3ab^2 = 12a^3b^5$

The numbers are multiplied and the indices are added for each letter. Note that a is the same as a^1, so $a^2 \times a = a^{2+1} = a^3$.

b) $\dfrac{12ab^3 \times 3a^2b}{2a^3b^2} = 18b^2$

The terms combine as
$12 \times 3 \div 2 = 18$,
$a \times a^2 \div a^3 = a^{1+2-3} = a^0 = 1$,
$b^3 \times b \div b^2 = b^{3+1-2} = b^2$.

c) $4a^2 + 3a^3$

This cannot be simplified. The two terms are not like terms and cannot be added.

STAGE
9

EXERCISE 2.2

Simplify where possible.

1 $3a^2 \times 4a^3$

2 $\dfrac{a^5 \times a^3}{a^6}$

3 $\dfrac{12a^5}{6a^3}$

4 $3a^2 \times 4a^2$

5 $(3a^3)^2$

6 $(2c)^3$

7 $2a^2b \times 3a^3b^2$

8 $3a^2b^3 \times 2a^3b^4$

9 $4a^2b - 2ab^2$

10 $12a^2 \times 3b^2$

11 $\dfrac{15a^2b^3 \times 3a^2b}{9a^3b^2}$

12 $2a^2 + 3a^3$

13 $\dfrac{9p^2q \times (2p^3q)^2}{12p^5q^3}$

14 $8a^2b^3 \times 2a^3b \div 4a^4b^2$

15 $\dfrac{4abc \times 3a^2bc^2}{6a^2bc^2}$

16 $\dfrac{(3a^2b^2)^3}{(a^3b)^2}$

17 $\dfrac{12t^3}{(2t)^2}$

18 $4a^2 \times 2b^3 - a \times 3b \times ab^2$

19 $2a^2b \times 3ab^2 - 4a^3b^3$

20 $6a^2 \times (2ab^2)^2 \div 12b^2a$

21 $\dfrac{6a^4b}{5c^2} \times \dfrac{10c^3}{9a^2b^3}$

22 $\dfrac{12x^5y^4}{7z^5} \times \dfrac{14z^3}{15x^3y^2}$

23 $\dfrac{6a^4b^5}{9c^3} \div \dfrac{10a^2b^3}{3c^6}$

24 $\dfrac{8a^7b^5}{15c^4} \div \dfrac{4ab^2}{9c}$

25 $\dfrac{6t^5v^2}{5w^2} \div \dfrac{8t^2v^5}{15w^4}$

26 $\dfrac{12e^6f^2}{5g^3} \times \dfrac{10g^4}{8e^4f^3}$

27 $\dfrac{5a^3b^2}{2c^4} \times \dfrac{8c^2d^4}{15ae^3} \times \dfrac{3e^2}{2b^3d^2}$

28 $\dfrac{8t^3v^2}{3x^5z^4} \times \dfrac{5x^2}{12v^4y^3} \div \dfrac{10t^4}{9y^2z^5}$

29 $\left(\dfrac{4a^5b^6}{5c^2d^4} \div \dfrac{12a^2b^5}{25c^4}\right) \times \dfrac{8d^2}{15ab^3}$

30 $(5x^2y^3 \div 2y^4) + 5x^3$

C CHALLENGE 3

Work in pairs.

Write your own algebraic fraction problem. Use either multiplication or division, but not both.

Challenge your partner to simplify the problem to a single algebraic fraction.

Factorising algebraic expressions

Factors are numbers or letters which will divide into an expression.

The factors of 6 are 1, 2, 3 and 6.
The factors of b^3 are 1, b, b^2 and b^3.

Remember that multiplying or dividing by 1 leaves a number unchanged, so 1 is not a useful factor and is often ignored.

To factorise an expression, look for **common factors**. For example, the common factors of $2a^2$ and $6a$ are 2, a and $2a$.

EXAMPLE 5

Factorise these fully.

a) $4p + 6$

b) $2a^2 - 3a$

c) $15ab^2 + 10a^2b^2$

d) $2a - 10a^2 + 6a^3$

a) $4p + 6 = 2(2p + 3)$ The only common factor is 2.
$2 \times 2p = 4p, 2 \times 3 = 6$

b) $2a^2 - 3a = a(2a - 3)$ The only common factor is a.
$a \times 2a = 2a^2, a \times {}^-3 = {}^-3a$

c) $15ab^2 + 10a^2b^2 = 5ab^2(3 + 2a)$ 5, a and b^2 are common factors.
$5ab^2 \times 3 = 15ab^2, 5ab^2 \times 2a = 10a^2b^2$

d) $2a - 10a^2 + 6a^3 = 2a(1 - 5a + 3a^2)$ 2 and a are common factors.
$2a \times 1 = 2a, 2a \times {}^-5a = {}^-10a^2$ and $2a \times 3a^2 = 6a^3$

EXAM TIP
Make sure that you have found all the common factors. Check that the expression in the bracket will not factorise further.

STAGE
9

EXERCISE 2.3

Factorise these fully.

1	$2a + 8$	**16**	$14a^2 - 8a^3$
2	$3x - 12$	**17**	$4a^2c - 2ac^2$
3	$3a + 5a^2$	**18**	$21x^2 - 14y^2$
4	$4a + 5ab$	**19**	$15xy - 5y$
5	$2ab - 6ac$	**20**	$12x^2y + 8xy - 4xy^2$
6	$4ab - 2a^2$	**21**	$6a^3 - 4a^2 + 2a$
7	$5a^2b + 10ab^2$	**22**	$14s^2t - 7st^2$
8	$3ab - 2ac + 3ad$	**23**	$3a^2b - 9a^3b^2$
9	$2x^2y^2 - 3x^3y$	**24**	$10z^3 - 25z^2 + 5z$
10	$5x^2 - 15x + 15$	**25**	$5a^2b^2c^2 - 10abc$
11	$3a^2b - 6ab^2$	**26**	$5abc - 15a^2b^2c^2$
12	$4a^2b - 3ab^2$	**27**	$2a^2b - 3a^2b^3 + 7a^4b$
13	$12x - 6y + 8z$	**28**	$3a^2bc - 6ab^2c - 9abc^2$
14	$9x^2y - 6xy^2$	**29**	$4abc - 3ac^2 + 2a^2b$
15	$9ab + 6b^2$	**30**	$7a^3b^3c^2 - 14a^2b^3c^3$

The difference of two squares

You saw on page 6 that expressions of the type $x^2 - b^2$ can be factorised to give $(x - b)(x + b)$.

(Check that these are the same by multiplying out the brackets and simplifying your answer.)

This method of factorising is important when there is no x-term in a quadratic expression.

EXAMPLE 6

Factorise $x^2 - 16$.

$$x^2 - 16 = x^2 - 4^2$$
$$= (x - 4)(x + 4)$$

In fact *any* expression which can be written as two squares subtracted can be factorised in this way.

EXAMPLE 7

Factorise each of these expressions.

a) $25x^2 - 1$

b) $9x^2 - 4y^2$

a) $25x^2 - 1 = 5^2x^2 - 1$
$$= (5x - 1)(5x + 1)$$

b) $9x^2 - 4y^2 = 3^2x^2 - 2^2y^2$
$$= (3x - 2y)(3x + 2y)$$

When the 'number term' is not a square number, check to see whether a common factor can be extracted, leaving an expression in a form that allows you to use the 'difference of two squares' method.

EXAMPLE 8

Factorise $3x^2 - 12$.

$$3x^2 - 12 = 3(x^2 - 4)$$ Take out the common factor.
$$= 3(x^2 - 2^2)$$
$$= 3(x - 2)(x + 2)$$

EXERCISE 2.4

Factorise each of these expressions.

1 $x^2 - 25$

2 $x^2 - 4$

3 $4a^2 - b^2$

4 $9 - 16y^2$

5 $25x^2 - 49y^2$

6 $9x^2 - 64$

7 $a^2 - 9b^2$

8 $1 - 49t^2$

9 $100x^2 - 1$

10 $25 - 4x^2$

11 $x^2y^2 - 16a^2$

12 $y^2 - 169$

13 $121x^2 - 144y^2$

14 $81p^2 - 36q^2$

15 $8 - 2x^2$

16 $3x^2 - 192$

17 $7a^2 - 63b^2$

18 $45 - 20x^2$

19 $25x^2y^2 - 100$

20 $x^2y^2z^2 - 100$

21 $3x^2 - 12$

22 $5x^2 - 45$

23 $3x^2 - 108$

24 $7x^2 - 343$

25 $10x^2 - 4000$

26 $8x^2 - 200$

Factorising quadratic expressions where the coefficient of $x^2 \neq 1$

You have already learned how to factorise simple quadratic expressions such as $x^2 + bx + c$.

Remember
- if c is positive find two numbers that multiply to c and add up to b.
- if c is negative find two numbers that multiply to c and whose difference is b.

The expression $ax^2 + bx + c$, when factorised, will be $(px + q)(rx + s) = prx^2 + (ps + qr)x + qs$ when expanded.

So $pr = a$, $qs = c$ and $ps + qr = b$. As before, the sign of c determines whether you are looking for a sum or a difference.

It is easiest to look at examples.

EXAMPLE 9

Factorise $3x^2 + 11x + 6$.

As the last sign is +, both signs in the bracket are the same, and as the middle sign is +, they are both +.

The only numbers that can multiply to give 3 are 3 and 1.

So, as a start, $(3x + ...)(x + ...)$.

The numbers that multiply to give 6 are either 3 and 2 or 6 and 1.

So the possible answers are

$(3x + 2)(x + 3)$ or $(3x + 3)(x + 2)$ or $(3x + 6)(x + 1)$ or $(3x + 1)(x + 6)$.

By expanding the brackets it can be seen that the first one is correct.

$3x^2 + 11x + 6 = (3x + 2)(x + 3)$

The coefficient of the middle term is $3 \times 3 + 2 \times 1 = 9 + 2 = 11$.
It is very useful to check completely by multiplying out the two brackets fully.

Writing out all the possible brackets can be a long process and it is quicker to test the possibilities for the middle term until you find the correct one and then multiply out the brackets to check.

EXAMPLE 10

Factorise $4x^2 - 14x + 6$.

First, check whether there is a common factor.

Here the common factor is 2.

$4x^2 - 14x + 6 = 2(2x^2 - 7x + 3)$

Now look at the quadratic expression.

Here you can see that both signs in the brackets are –.

The possibilities for the coefficients of the first terms in the brackets are 2 and 1.
The possibilities for the coefficients of the second terms are 3 and 1 or 1 and 3.

The coefficient of the middle term is thus $2 \times 1 + 1 \times 3 = 5$ or $2 \times 3 + 1 \times 1 = 7$.
The second is correct so $2x^2 - 7x + 3 = (2x - 1)(x - 3)$ and the full answer is
$4x^2 - 14x + 6 = 2(2x - 1)(x - 3)$.

STAGE
9

15

2

EXAMPLE 11

Factorise $5x^2 + 13x + 6$.

There is no common factor.

Both signs in the brackets are +.

The coefficients of the first terms are 5 and 1; the coefficients of the second terms are 1 and 6, 6 and 1, 3 and 2, or 2 and 3.

Test the possibilities for the middle term until you find the correct one.
$5 \times 1 + 1 \times 6 = 11$, $5 \times 6 + 1 \times 1 = 31$, $5 \times 3 + 1 \times 2 = 17$, $5 \times 2 + 1 \times 3 = 13$.

So $5x^2 + 13x + 6 = (5x + 3)(x + 2)$. Check by multiplying out.

EXAMPLE 12

Factorise $6x^2 - 17x + 12$.

There is no common factor.

Both signs in the brackets are −.

The coefficients of the first terms are 6 and 1 or 3 and 2; the coefficients of the second terms are 1 and 12, 12 and 1, 2 and 6, 6 and 2, 4 and 3 or 3 and 4.

This could mean 12 possible products but if you look at the middle term and see that the coefficient is 17 you can gather that you will not be multiplying anything by 12 and are unlikely to multiply anything by 6. Try the most likely ones first.

$3 \times 4 + 2 \times 3 = 18$, $3 \times 3 + 2 \times 4 = 17$ which is correct.

So $6x^2 - 17x + 12 = (3x - 4)(2x - 3)$. Check by multiplying out.

STAGE

9

EXAM TIP

First look for any common factor, then try the most obvious pairs first and remember, if the sign of c is positive, everything is added and both brackets have the same sign as b.

EXERCISE 2.5

Factorise each of these expressions.

1 $x^2 + 7x + 6$

2 $x^2 + 5x + 6$

3 $x^2 - 6x + 8$

4 $x^2 - 7x + 10$

5 $2x^2 + 6x + 4$

6 $3x^2 + 7x + 2$

7 $2x^2 + 9x + 4$

8 $2x^2 + 7x + 6$

9 $6x^2 - 15x + 6$

10 $3x^2 - 12x + 12$

11 $3x^2 - 11x + 6$

12 $3x^2 - 13x + 10$

13 $3x^2 - 11x + 10$

14 $4x^2 - 16x + 15$

15 $4x^2 + 8x + 3$

16 $7x^2 + 10x + 3$

17 $5x^2 - 13x + 6$

18 $5x^2 - 22x + 8$

19 $6x^2 - 19x + 10$

20 $8x^2 - 18x + 9$

21 $3x^2 + 17x + 20$

22 $2x^2 + 7x + 6$

23 $3x^2 + 13x + 4$

24 $5x^2 + 18x + 9$

25 $4x^2 + 6x + 2$

26 $3x^2 + 11x + 10$

27 $2x^2 + 5x + 2$

28 $4x^2 + 17x + 15$

C CHALLENGE 4

Factorise each of these expressions.

a) $8x^2 + 10x + 3$

b) $15x^2 + 2x - 8$

c) $8x^2 - 2x - 15$

d) $6x^2 - 29x + 35$

Factorising quadratic expressions with a negative constant term

The examples looked at so far had positive c and so were straightforward. When c is negative the signs in the brackets are different and the middle term is the difference of the products.

These are, again, best shown by examples.

EXAMPLE 13

Factorise $3x^2 - 7x - 6$.

There is no common factor.

The signs in the brackets are different.

The coefficients of the first terms are 3 and 1; the coefficients of the second terms are 1 and 6, 6 and 1, 3 and 2 or 2 and 3.

Test the possibilities for the middle term until you find the correct one. $3 \times 1 - 1 \times 6 = {}^-3$, $3 \times 6 - 1 \times 1 = 17$, $3 \times 3 - 1 \times 2 = 7$. This is the correct number but the wrong sign. So the 3×3 must be negative, not the 1×2, which means that the second term in one of the brackets must be $^-3$.

So $3x^2 - 7x - 6 = (3x + 2)(x - 3)$. As there was no common factor the 3s must be in separate brackets.

Multiply out to check.

EXAMPLE 14

Factorise $6x^2 + 3x - 30$.

The common factor is 3.

$6x^2 + 3x - 30 = 3(2x^2 + x - 10)$

Look at the quadratic expression. The signs in the brackets are different.

The coefficients of the first terms are 2 and 1; the coefficients of the second terms are 1 and 10, 10 and 1, 5 and 2 or 2 and 5.

Try the easiest products first.

$2 \times 5 - 1 \times 2 = 8$, $2 \times 2 - 1 \times 5 = -1$ which is the correct number but the wrong sign.

So the second term in one of the brackets is $^-2$.

So $6x^2 + 3x - 30 = 3(2x + 5)(x - 2)$.

Check by multiplying out.

STAGE 9

Algebraic manipulation

EXAMPLE 15

Factorise $6x^2 - 5x - 4$.

There is no common factor.

The signs in the brackets are different.

The coefficients of the first terms are 6 and 1 or 3 and 2; the coefficients of the second terms are 4 and 1, 1 and 4 or 2 and 2.

Try the easiest products first.

$3 \times 2 - 2 \times 2 = 2$, $3 \times 4 - 2 \times 1 = 10$,
$3 \times 1 - 2 \times 4 = {}^-5$ which is correct.

So $6x^2 - 5x - 4 = (3x - 4)(2x + 1)$.

Check by multiplying out.

> ### EXAM TIP
> If the sign of c is negative, find differences of products, then put the correct numbers in the brackets before deciding on the signs. Always check by multiplying out.

EXERCISE 2.6

Factorise each of these expressions.

1	$x^2 - x - 6$	**15**	$3x^2 - x - 14$
2	$x^2 - 3x - 18$	**16**	$3x^2 - 11x - 20$
3	$x^2 + 3x - 10$	**17**	$2x^2 - x - 21$
4	$3x^2 + x - 10$	**18**	$2x^2 - 15x - 8$
5	$2x^2 + 5x - 3$	**19**	$6x^2 - 17x - 14$
6	$2x^2 - 18$	**20**	$6x^2 - 13x - 15$
7	$3x^2 - 2x - 8$	**21**	$2x^2 - x - 15$
8	$3x^2 - 11x - 4$	**22**	$3x^2 + x - 14$
9	$2x^2 + 9x - 5$	**23**	$5x^2 - 17x - 12$
10	$3x^2 + 4x - 15$	**24**	$3x^2 - 5x - 12$
11	$5x^2 - 15x - 50$	**25**	$4x^2 - 3x - 10$
12	$5x^2 + 13x - 6$	**26**	$2x^2 - 7x - 15$
13	$4x^2 - 4x - 3$	**27**	$4x^2 - 7x - 2$
14	$7x^2 + 10x - 8$	**28**	$3x^2 - 16x - 12$

STAGE
9

Cancelling algebraic fractions

When cancelling fractions, it is factors that cancel, never part of factors.

EXAMPLE 16

Simplify $\dfrac{4ab^2}{3c^2} \times \dfrac{9c^2}{2a^2b}$.

$$\frac{4ab^2}{3c^2} \times \frac{9c^2}{2a^2b} = \frac{6b}{a}$$

2, 3, a, b and c^2 all cancel.

EXAMPLE 17

Simplify $\dfrac{x^2 + x}{x^2 - 2x - 3}$.

As it stands this fraction cannot be cancelled. First both numerator and denominator must be factorised.

$$\frac{x^2 + x}{x^2 - 2x - 3} = \frac{x(x + 1)}{(x - 3)(x + 1)}$$

$$= \frac{x}{(x - 3)}$$

$(x + 1)$ cancels.

EXAM TIP

Errors often occur by cancelling individual terms. Only factors, which can be individual numbers, letters or brackets, can be cancelled.

EXERCISE 2.7

Simplify each of these expressions.

1 $\dfrac{15a^2}{6} \times \dfrac{b^2}{a}$

2 $\dfrac{12abc}{a^2b} \times \dfrac{a^3b}{4c}$

3 $\dfrac{x^3y^2}{10x} \times \dfrac{15xy}{10y^2}$

4 $\dfrac{2x^2y^2}{xy} \times \dfrac{3xy^3}{4x^2}$

5 $\dfrac{2x}{x^2 - 3x}$

6 $\dfrac{5x^2 - 20x}{10x^2}$

7 $\dfrac{3x^2 - 6x}{x^2 + x - 6}$

8 $\dfrac{x^2 + 2x + 1}{x^2 - 1}$

9 $\dfrac{x^2 - 5x + 4}{x^2 - 2x - 8}$

10 $\dfrac{3x^2 + 5x - 2}{x^2 + 7x + 10}$

EXERCISE 2.7 continued

11 $\dfrac{x^2 - 5x + 6}{x^2 - 9}$

12 $\dfrac{x^2 - 16}{3x^2 + 12x}$

13 $\dfrac{2x^2 + 3x - 5}{4x^2 - 25}$

14 $\dfrac{3x^2 + 3x - 18}{2x^2 - 18}$

15 $\dfrac{5x^2 - 80}{2x^2 - 8x}$

16 $\dfrac{3x^2 - x - 4}{5x^2 - 5}$

17 $\dfrac{3x + 15}{x^2 + 3x - 10}$

18 $\dfrac{6x - 18}{x^2 - x - 6}$

19 $\dfrac{x^2 - 5x + 6}{x^2 - 4x + 3}$

20 $\dfrac{x^2 - 3x - 4}{x^2 - 4x - 5}$

21 $\dfrac{x^2 - 2x - 3}{x^2 - 9}$

22 $\dfrac{3x^2 - 12}{x^2 + 2x - 8}$

23 $\dfrac{3x^2 + 5x + 2}{2x^2 - x - 3}$

24 $\dfrac{2x^2 + x - 6}{x^2 + x - 2}$

25 $\dfrac{6x^2 - 3x}{(2x - 1)^2}$

26 $\dfrac{5(x + 3)^2}{x^2 - 9}$

27 $\dfrac{(x - 3)(x + 2)^2}{x^2 - x - 6}$

28 $\dfrac{2x^2 + x - 6}{(2x - 3)^2}$

Solving quadratic equations

In Stage 8 you learned how to solve quadratic equations by writing the equation in the form $x^2 + bx + c = 0$, factorising the left-hand side and equating each of the factors in turn to zero. Earlier in this chapter you learned how to factorise expressions of the form $ax^2 + bx + c$. By equating each of the factors in turn to zero, you can now solve equations of the form $ax^2 + bx + c = 0$ as well.

EXAMPLE 18

Solve the equation $3x^2 + 11x + 6 = 0$.

This is the expression factorised in Example 9.

$3x^2 + 11x + 6 = 0$

$(3x + 2)(x + 3) = 0$

$\qquad (3x + 2) = 0 \quad$ or $\quad (x + 3) = 0$

$\qquad\qquad 3x = {}^-2 \quad$ or $\qquad x = {}^-3$

$\qquad\qquad x = \dfrac{{}^-2}{3} \quad$ or $\qquad x = {}^-3$

EXAMPLE 19

Solve the equation $4x^2 - 14x + 6 = 0$.

This is the expression factorised in Example 10.

It is factorised by taking out a common factor.

$$4x^2 - 14x + 6 = 0$$
$$2(2x^2 - 7x + 3) = 0$$
$$2(2x - 1)(x - 3) = 0$$

Divide both sides by the common factor, 2. (Remember that zero divided by anything is zero.)

$$(2x - 1)(x - 3) = 0$$

$(2x - 1) = 0$	or	$(x - 3) = 0$
$2x = 1$	or	$x = 3$
$x = \frac{1}{2}$	or	$x = 3$

EXAMPLE 20

Solve $2x^2 - 32 = 0$.

As before, first look for a common factor. Here it is 2.

$$2x^2 - 32 = 0$$
$$2(x^2 - 16) = 0$$

You should recognise that $(x^2 - 16)$ is the difference of two squares.

$$2(x + 4)(x - 4) = 0$$
So $\qquad x = \pm 4$

EXERCISE 2.8

Solve these equations.

1 $2x^2 - 5x - 12 = 0$		**4** $2x^2 - 3x - 5 = 0$		**7** $2x^2 - 13x + 15 = 0$
2 $3x^2 - 10x - 8 = 0$		**5** $3x^2 + 2x - 1 = 0$		**8** $12x^2 + 10x - 8 = 0$
3 $2x^2 + 5x + 3 = 0$		**6** $2x^2 + 11x + 5 = 0$		**9** $2x^2 + 2x - 60 = 0$

EXERCISE 2.8 continued

10 $3x^2 - 12x + 9 = 0$

11 $2x^2 - 2x - 12 = 0$

12 $3x^2 - 14x - 24 = 0$

13 $2x^2 - 8 = 0$

14 $3x^2 - 27 = 0$

15 $5x^2 - 125 = 0$

16 $2x^2 - 72 = 0$

17 $6x^2 - 54 = 0$

18 $10x^2 - 1000 = 0$

 CHALLENGE 5

$S = \frac{1}{2}n(n + 1)$ gives the sum, S, of the first n positive integers.

Find n when $S = 325$.

 KEY IDEAS

- When multiplying two brackets, multiply every term in the first bracket by every term in the second bracket.

- When multiplying algebraic expressions involving powers, add the indices.

- When dividing algebraic expressions involving powers, subtract the indices.

- An expression of the type $x^2 - b^2$ can be factorised as $(x - b)(x + b)$. This is called the difference of two squares.

- When factorising, first look for common factors.

- When factorising expressions of the form $ax^2 + bx + c$ find numbers to fit

  $(px + q)(rx + s)$ such that $pr = a$, $qs = c$ and $ps + qr = b$.

  If c is positive, it is a sum and if c is negative, it is a difference.

- Check you have factorised an expression correctly by multiplying out the brackets.

- When cancelling algebraic fractions, factorise if necessary and then cancel factors.

- Solve quadratic equations of the form $ax^2 + bx + c = 0$ by factorising the left-hand side and equating each of the factors in turn to zero.

STAGE
9

3 Proportion and variation

Proportion

You have met proportion before. Look at this example.

EXAMPLE 1

A car uses 12 litres of petrol to travel 100 km.

How many litres will it use to travel 250 km?

This is an example of **direct proportion**. As the distance increases, so does the amount of petrol used.

The distance has been increased by multiplying by a factor of $\frac{250}{100} = 2.5$.

Increase the amount of petrol by multiplying by the same factor.
Amount of petrol = $12 \times 2.5 = 30$ litres.

In Example 1, if the distance is x km and the amount of petrol is y litres, then y is directly proportional to x. In other words, y varies in the same way as x. If you double x, you also double y. It is possible to express this in symbols as $y \propto x$, which is read as 'y varies as x' or 'y is proportional to x'.

Now look at this example.

EXAMPLE 2

It takes 4 men 10 days to dig a hole.

How long will it take 20 men?

This is an example of **inverse proportion**. As the number of men increases, the time taken will decrease.

The number of men has increased by multiplying by a factor of $\frac{20}{4} = 5$.

Decrease the time by dividing by the same factor.

Number of days = $10 \div 5 = 2$.

(Of course, this does assume that there will be room for all the men in the hole!)

This time, if the number of men is x and the number of days is y, then y is inversely proportional to x. In other words, y varies inversely as x. If you double x, you halve y. This is the same as 'y varies as $\frac{1}{x}$', which is written in symbols as $y \propto \frac{1}{x}$.

EXAM TIP

You can easily tell whether the proportion is direct or inverse – in direct proportion, both variables change in the same way, either up or down; in inverse proportion, when one variable goes up, the other will go down.

EXERCISE 3.1

1   Describe the variation in each of these situations, using the symbol $\propto$.
  a) The length of tape, y, and the time of the recording, x.
  b) The cost of a train ticket, y, and the length of the journey, x.
  c) The time the journey takes, t, and the speed of the train, s.
  d) The number of pages in a book, p, and the number of words, w.
  e) The probability my ticket wins the raffle, p, and the number of raffle tickets sold, n.

  f) The depth of water in a rectangular tank, d, and the length of time it has been filling, t.
  g) The number of buses, b, needed to carry 2000 people and the number of seats on a bus, s.
  h) The time a journey takes, t, at a fixed speed and the distance covered, d.
  i) The number of ice creams you can buy, c, and the amount of money you have, m.
  j) The probability I win the raffle, p, and the number of raffle tickets I buy, n.

STAGE
9

EXERCISE 3.1 continued

2 Describe the variation shown in each of these tables of values. Use the symbol ∝.

a)

x	3	15
y	1	5

f)

x	3	15
y	5	1

b)

x	4	15
y	28	105

g)

x	3	15
y	2	10

c)

x	8	20
y	10	4

h)

x	8	20
y	10	25

d)

x	10	12
y	50	60

i)

x	1	0·1
y	50	500

e)

x	24	4·8
y	16	3·2

j)

x	16	56
y	6·4	22·4

Variation as a formula

If $y \propto x$ then the same factor is applied to y as was applied to x.

Look at the table on the right.

x	5	15
y	3	9

We can tell that $y \propto x$ as $5 \times 3 = 15$ for x and $3 \times 3 = 9$ for y.

Now look at the pairs of values of x and y, (5, 3) and (15, 9). In the first pair, $\frac{y}{x} = \frac{3}{5}$.

But this is also true for the second pair, as $\frac{9}{15} = \frac{3}{5}$.

Rewriting the equation $\frac{y}{x} = \frac{3}{5}$ gives the formula $y = \frac{3}{5}x$.

Another way to look at the problem is to graph the equation $y = \frac{3}{5}x$.

Check that the points (5, 3) and (15, 9) are on the graph.

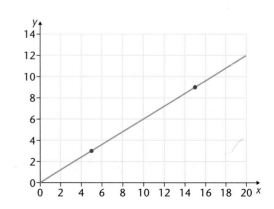

EXAMPLE 3

Find the formula for Example 2.

x	4	20
y	10	2

This time $y \propto \dfrac{1}{x}$.

Write the values of $\dfrac{1}{x}$ in a table.

$\dfrac{1}{x}$	0·25	0·05
y	10	2

So now $y \div \dfrac{1}{x} = 10 \div 0\cdot25$

Hence $y \div \dfrac{1}{x} = 40$.

This gives the formula $y = \dfrac{40}{x}$.

Check that it works in the table.

There is another way to do this, which you may have spotted already.

In the first table, find the values of $x \times y$. This is 40, so the formula is $xy = 40$, which is equivalent to $y = \dfrac{40}{x}$.

Here is the graph of $y = \dfrac{40}{x}$.

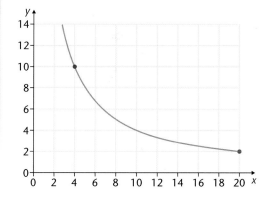

Check that the points (4, 10) and (20, 2) are on the graph.

EXERCISE 3.2

1   Find a formula for each variation in Exercise 3.1, question **2**.
    In each case, sketch the graph.

2   Ohm's law states that if a voltage, V volts, is applied across the terminals of an electrical component, and if the current flowing in the component is I amperes, then $I \propto V$.

    If $I = 2 \cdot 5$ when $V = 15$

    a) find the formula for the variation.
    b) find I when $V = 60$ volts.

3   If w m is the wavelength of a sound wave and f hertz is the frequency of the wave, then $w \propto \dfrac{1}{f}$.

    If $w = 1 \cdot 1$ m when $f = 300$ hertz

    a) find the formula for the variation.
    b) find w when $f = 660$ hertz.

Other types of variation

Sometimes two variables can be related in more complicated ways.

Here you will meet $y \propto x^2$, $y \propto x^3$ and $y \propto \dfrac{1}{x^2}$.

EXAMPLE 4

$y \propto x^2$.

If $y = 10$ when $x = 5$, what is y when $x = 15$?

x has been multiplied by $15 \div 5 = 3$,
so y will be multiplied by $3^2 = 9$.

$y = 10 \times 9 = 90$

EXAMPLE 5

$y \propto \dfrac{1}{x^2}$.

> This is sometimes called the **inverse square law**, for obvious reasons.

If $y = 10$ when $x = 5$, what is y when $x = 10$?

x has been multiplied by $10 \div 5 = 2$,
so y will be divided by $2^2 = 4$.

$y = 10 \div 4 = 2 \cdot 5$

EXAM TIP

When finding the variation or the formula, start by deciding whether the proportion is direct or inverse. This reduces the number of possibilities to be tried.

EXERCISE 3.3

1 $y \propto x^2$ and $y = 3$ when $x = 6$.
Find y when $x = 12$.

2 $y \propto x^2$ and $y = 9$ when $x = 7.5$.
Find y when $x = 5$.

3 $y \propto x^3$ and $y = 1$ when $x = 3$.
Find y when $x = 6$.

4 $y \propto \dfrac{1}{x^2}$ and $y = 4$ when $x = 4$.
Find y when $x = 8$.

5 $y \propto x^2$ and $y = 5$ when $x = 6$.
Find y when $x = 3$.

6 $y \propto \dfrac{1}{x^2}$ and $y = 10$ when $x = 3$.
Find y when $x = 12$.

7 $y \propto \dfrac{1}{x^2}$ and $y = 10$ when $x = 4$.
Find y when $x = 6$.

8 $y \propto x^3$ and $y = 12$ when $x = 5$.
Find y when $x = 10$.

9 $y \propto x^3$ and $y = 3$ when $x = 10$.
Find y when $x = 5$.

10 $y \propto x^2$ and $y = 7$ when $x = 8$.
Find y when $x = 6$.

11 $y \propto x^3$ and $y = 4$ when $x = 5$.
Find y when $x = 10$.

12 $y \propto x^2$ and $y = 2$ when $x = 2$.
Find y when $x = 8$.

13 $y \propto \dfrac{1}{x^2}$ and $y = 7$ when $x = 7$.
Find y when $x = 14$.

14 $y \propto \dfrac{1}{x}$ and $y = 1$ when $x = 1$.
Find y when $x = 0.5$.

15 $y \propto x$ and $y = 8$ when $x = 3$.
Find y when $x = 10.5$.

16 $y \propto \dfrac{1}{x^2}$ and $y = 3$ when $x = 1$.
Find y when $x = 0.5$

17 $y \propto x^3$ and $y = 14$ when $x = 12$.
Find y when $x = 15$.

18 $y \propto x^2$ and $y = 2$ when $x = 6.5$.
Find y when $x = 19.5$.

19 $y \propto x^2$ and $y = 5$ when $x = 0.6$.
Find y when $x = 2.4$.

20 $y \propto \dfrac{1}{x^2}$ and $y = 9$ when $x = 3$.
Find y when $x = 1$.

21 Describe the variation shown in each of these tables of values.
Use the symbol $\propto$.

a)

x	5	25
y	5	125

b)

x	5	10
y	5	1.25

c)

x	5	15
y	5	135

d)

x	5	2.5
y	5	10

e)

x	4	6
y	18	8

f)

x	2	6
y	7	63

g)

x	1	0.25
y	1	4

h)

x	54	21.6
y	33	13.2

i)

x	16	8
y	15	1.875

j)

x	24	48
y	4	1

Finding formulae for other types of variation

Finding formulae for other variations can be done in a similar way to the previous ones.

This is another approach, however.

Here is the result of Example 4 in a table.

x	5	15
y	10	90

$y \propto x^2$ so try the formula $y = kx^2$, where k is a constant to be found.

Substitute 5 and 10 for x and y.

$10 = k \times 5^2$, giving $k = 0\cdot4$.

Substitute 15 and 90 as a check.

$90 = 0\cdot4 \times 15^2$, which is correct.

So $y = 0\cdot4x^2$.

Here is the graph of $y = 0\cdot4x^2$.

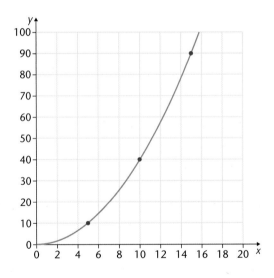

Check that the points from the table are on the graph.

Here is the result of Example 5 in a table.

x	5	10
y	10	2·5

$y \propto \dfrac{1}{x^2}$, so try the formula $y = \dfrac{k}{x^2}$.

Substitute 5 and 10 for x and y.

$10 = \dfrac{k}{5^2}$, giving $k = 250$.

Substitute 10 and 2·5 as a check.

$2·5 = \dfrac{250}{100}$, which is correct.

So $y = \dfrac{250}{x^2}$.

Here is the graph of $y = \dfrac{250}{x^2}$.

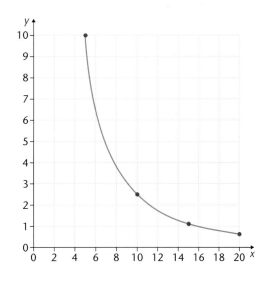

Check that the points from the table are on the graph.

EXERCISE 3.4

Find the formula for each of the questions in Exercise 3.3 and sketch the graph.

EXERCISE 3.5

1   For each of the relationships shown in the tables below and overleaf,
    (i) state the type of proportion.
    (ii) find the formula.
    (iii) where appropriate, find the missing y value in the table.

a)

x	4	6
y	16	36

b)

x	2	5	8
y	8	50	

EXERCISE 3.5 continued

c)

x	6	9
y	12	27

d)

x	5	25	35
y	10	250	

e)

x	3	6
y	10·8	43·2

f)

x	3	6	15
y	25	6·25	

g)

x	5	10
y	50	12·5

h)

x	2	5	10
y	5	0·8	

i)

x	10	20
y	8	2

j)

x	8	10	16
y	4	2·56	

2 The wind resistance of a train is proportional to the square of the train's speed.
  a) If the speed is doubled, what happens to the wind resistance?
  b) What happens to the resistance when the speed is multiplied by 1·5?

3 If a ball is thrown upwards with a speed of u m/s and rises to a maximum height of h metres, then $h \propto u^2$. If $h = 90$ when $u = 42$
  a) find the formula for the variation.
  b) calculate h when $u = 65$.

4 The surface area, S, of a sphere is proportional to the square of the diameter, d. Assume that planets are perfect spheres, and that the diameter of Mars is approximately twice the diameter of the Moon and the surface area of the Moon is approximately 3.8×10^7 km^2.
  Calculate the approximate surface area of Mars.

C CHALLENGE 1

The mass of copper wire depends jointly upon its length and the square of its diameter.

If 500 m of wire with a diameter of 3 mm has a mass of 31.5 kg, what will be the mass of 1 km of wire with a diameter of 2 mm.

K KEY IDEAS

- The symbol '$\propto$' means 'varies as' or 'is proportional to'.

- Direct proportion includes $y \propto x$, $y \propto x^2$ and $y \propto x^3$.
  The formulae for these are $y = kx$, $y = kx^2$ and $y = kx^3$, respectively.

- Inverse proportion includes $y \propto \dfrac{1}{x}$ and $y \propto \dfrac{1}{x^2}$.

  The formulae for these are $y = \dfrac{k}{x}$ and $y = \dfrac{k}{x^2}$, respectively.

Revision exercise A1

1 Find an approximate answer to each of these calculations by rounding each number to 1 significant figure.
   a) 580×83
   b) $\dfrac{63\cdot2}{3\cdot8}$
   c) $\dfrac{28\cdot3^2}{0\cdot48}$

   Now use a calculator to see how close your approximations are to the correct answers.

2 Simplify each of these, writing the answers in standard form.
   a) $5\cdot4 \times 10^6 \times 2\cdot7 \times 10^{-3}$
   b) $9\cdot54 \times 10^3 \div 4 \times 10^4$

3 Multiply out these brackets.
   a) $(x + 7)(x + 1)$
   b) $(a - 3)(a + 5)$
   c) $(2y - 4)(y + 1)$
   d) $(x - 5)(2x + 1)$
   e) $(4a + b)(a - b)$

4 Simplify these expressions.
   a) $2a^2 \times a^3$
   b) $10a^2 \div 2a$
   c) $(a^3)^2 \times a^3 \div a^4$
   d) $12a^2b \times 2a^2b^3$
   e) $6x^2y^2z^2 \div 2xy^2z$
   f) $8a^2b \times \dfrac{3abc}{6ab^2}$

5 Factorise each of these fully.
   a) $3a + 6b - 12c$
   b) $2a + 3ab$
   c) $a^2b - 3ab^2$
   d) $2x^2y - 6xy$
   e) $7abc + 14a^2b$
   f) $9a^2 + 3b^2 - 6c^2$
   g) $5pq - 10$
   h) $2a - 4a^2 + 6a^3$
   i) $100abc - 50ac$

6 Factorise these, where possible.
   a) $x^2 - 49$
   b) $x^2 - 1$
   c) $x^2 + 16$
   d) $x^2 - y^2$
   e) $121 - b^2$

7 Factorise these expressions.
   a) $x^2 - 16x + 63$
   b) $2x^2 - 8x - 42$
   c) $3x^2 - 8x + 4$
   d) $2x^2 + x - 15$
   e) $3x^2 - 48$
   f) $2x^2 - 11x - 21$
   g) $6x^2 - 27x - 15$
   h) $5x^2 - 21x + 18$
   i) $8x^2 - 6x - 5$
   j) $6x^2 - 11x - 10$

8 Simplify these fractions.
   a) $\dfrac{x^2 + 2x - 3}{4x - 4}$
   b) $\dfrac{x^2 + 2x + 1}{x^2 - x - 2}$
   c) $\dfrac{2x^2 + 2x}{x^2 + 4x + 3}$
   d) $\dfrac{x^2 - 7x + 10}{2x^2 - 11x + 5}$
   e) $\dfrac{x^2 + 8x + 15}{x^2 - 25}$
   f) $\dfrac{3x^2 + 15x + 12}{2x^2 + 7x - 4}$

9 Solve each of these equations.
   a) $3x^2 - 5x - 12 = 0$
   b) $2x^2 + x - 28 = 0$
   c) $3x^2 - 192 = 0$

10 The attractive force between two objects is inversely proportional to the square of their distance apart. The force is $0\cdot24$ units when the distance is 15 units.
   a) What will be the force when the distance is 30 units?
   b) Find a formula for the force, F, in terms of the distance apart, d.

11 Copy and complete the table to find the corresponding values of *y*.

	x	y
$y \propto x$	2	5
$y \propto x^2$	10	
$y \propto x^3$	10	
$y \propto \dfrac{1}{x}$	10	
$y \propto \dfrac{1}{x^2}$	10	

12 For each of these relationships
(i) find the variation, using $\propto$
(ii) find the formula.

a)

x	5	10
y	10	5

b)

x	5	50
y	10	100

c)

x	2	10
y	0·1	0·02

d)

x	2	10
y	0·1	2·5

Indices

You will learn about

- Using the laws of indices
- Dealing with negative and fractional indices

You should already know

- How to work out numbers with positive and negative indices
- How to use the basic laws of indices
- How to give answers to a number of significant figures
- The meaning of the words *power*, *prime* and *factor*

Fractional indices

You have already looked at numbers with positive and negative indices and should

know that $n^4 = n \times n \times n \times n$, $\quad n^{-2} = \dfrac{1}{n^2}$, $\quad \left(\dfrac{a}{b}\right)^{-1} = \dfrac{b}{a}$ $\quad$ and $\quad n^0 = 1$.

Suppose $a^b = \sqrt[3]{a}$.

Then $a^b \times a^b \times a^b = \left(\sqrt[3]{a}\right)^3 = a = a^1$ $\qquad$ Cube both sides.

$\qquad a^{b+b+b} = a^1$

$\qquad\quad a^{3b} = a^1$

$\qquad\quad 3b = 1$ $\qquad\qquad$ Equate powers.

$\qquad\quad b = \tfrac{1}{3}$

Therefore $a^{\frac{1}{3}} = \sqrt[3]{a}$.

A similar proof can be given to show that $a^{\frac{1}{2}} = \sqrt{a}$ and so on, and also $a^{\frac{1}{n}} = \sqrt[n]{a}$.

Similarly $a^{\frac{3}{2}} = \left(a^{\frac{1}{2}}\right)^3 = \left(\sqrt{a}\right)^3$ or $\left(a^3\right)^{\frac{1}{2}} = \sqrt{a^3}$.

STAGE
9

Indices

EXAMPLE 1

Write the following in index form, as simply as possible.

a) The cube of n b) $\dfrac{1}{n^3}$ c) $\sqrt[5]{n}$

a) The cube of $n = n^3$ b) $\dfrac{1}{n^3} = n^{-3}$ c) $\sqrt[5]{n} = n^{\frac{1}{5}}$

You will be asked to work out powers of numbers both with and without a calculator.

EXAMPLE 2

Write each of these as a whole number or a fraction. Do not use a calculator.

a) 3^2

b) $16^{\frac{1}{4}}$

c) $343^{\frac{1}{3}}$

d) 4^{-2}

e) $\left(\frac{1}{3}\right)^{-2}$

f) 6^0

g) $125^{\frac{2}{3}}$

a) $3^2 = 3 \times 3 = 9$

b) $16^{\frac{1}{4}} = 2$ $2 \times 2 \times 2 \times 2 = 16$

c) $343^{\frac{1}{3}} = 7$ $7 \times 7 \times 7 = 343$

d) $4^{-2} = \dfrac{1}{4^2} = \dfrac{1}{16}$

e) $\left(\frac{1}{3}\right)^{-2} = 3^2 = 9$

f) $6^0 = 1$ Any number raised to the power 0 is equal to 1.

g) $125^{\frac{2}{3}} = \left(\sqrt[3]{125}\right)^2 = 5^2 = 25$

STAGE
9

EXAM TIP
If you have to work out the square root of the cube of a number, it is usually easier to find the root first.

Working with a calculator

On your calculator you will find a key which will enable you to work out numbers such as $3 \cdot 1^5$.

It may be labelled $\boxed{\wedge}$. Find the powers key on your calculator. If you are not sure how to use it, practise with something simple like 2^4 which you can work out as 16. Try $\boxed{2}$ $\boxed{\wedge}$ $\boxed{4}$ $\boxed{=}$.

Similarly, your calculator will have a function which enables you to work out numbers like $2 \cdot 5^{\frac{1}{4}}$. It may be labelled $\sqrt[x]{}$ and be operated by pressing $\boxed{\text{SHIFT}}$ $\boxed{\wedge}$. Again practise with calculations you can work out in your head to check you are correct. You could do Example 2 again, this time using a calculator to check how your calculator works.

EXAMPLE 3

Use a calculator to work these out. Give your answers either exactly or to 5 significant figures.

a) $3 \cdot 5^4$

b) $2 \cdot 4^6$

c) $1 \cdot 03^{-3}$

d) $2 \cdot 15^{\frac{1}{4}}$

e) $3125^{\frac{4}{5}}$

a) $3 \cdot 5^4 = 150 \cdot 0625$
$\qquad = 150 \cdot 06$ (to 5 s.f.)

b) $2 \cdot 4^6 = 191 \cdot 102\,976$
$\qquad = 191 \cdot 10$ (to 5 s.f.)

c) $1 \cdot 03^{-3} = 0 \cdot 915\,141\,659$
$\qquad = 0 \cdot 915\,14$ (to 5 s.f.)

d) $2 \cdot 15^{\frac{1}{4}} = 1 \cdot 210\,903\,724$
$\qquad = 1 \cdot 2109$ (to 5 s.f.)

e) $3125^{\frac{4}{5}} = 625$

EXAM TIP

When finding a root by calculator it is easy to make a mistake. It is very helpful to check by working backwards.

For example, in part **d)** of Example 3, $2 \cdot 15^{\frac{1}{4}} = 1 \cdot 210\,903\,724$. Check: $1 \cdot 210\,903\,724^4 = 2 \cdot 149\,999\,997 \approx 2 \cdot 15$. ✓

Indices

STAGE
9

A ACTIVITY 1

Without using a calculator, match a number written in index form from the left-hand box to a whole number from the right-hand box.

2^3	$\left(\frac{1}{2}\right)^{-2}$	$1000^{\frac{2}{3}}$
$64^{\frac{1}{2}}$	10^2	$64^{\frac{1}{3}}$
$(0\cdot05)^{-1}$	$8^{\frac{2}{3}}$	$4^{\frac{1}{2}}$
$64^{\frac{2}{3}}$	$64^{\frac{1}{6}}$	7^0
$36^{\frac{1}{2}}$	$64^{\frac{5}{6}}$	

0	1	2
4	6	8
9	10	16
20	32	64
100	200	

EXERCISE 4.1

1 Write each of these in index form.
  a) The cube root of n **b)** The reciprocal of n^3
  c) $\sqrt[5]{n^2}$ **d)** $\sqrt[4]{n}$
  e) The reciprocal of $\left(\frac{1}{n}\right)^4$ **f)** $\sqrt[3]{n^5}$

Do not use your calculator for questions **2** to **11**.

Work out these. Give the answers as whole numbers or fractions.

2 a) 4^{-1} **b)** $4^{\frac{1}{2}}$ **c)** 4^0 **d)** 4^{-2} **e)** $4^{\frac{3}{2}}$

3 a) $8^{\frac{1}{3}}$ **b)** 8^{-1} **c)** $8^{\frac{4}{3}}$ **d)** $\left(\frac{1}{8}\right)^{-2}$ **e)** 8^1

4 a) 9^{-1} **b)** $9^{\frac{1}{2}}$ **c)** 9^0 **d)** 9^{-2} **e)** $9^{\frac{3}{2}}$

5 a) $27^{\frac{1}{3}}$ **b)** $27^{\frac{4}{3}}$ **c)** 27^{-1} **d)** $\left(\frac{1}{27}\right)^{-\frac{1}{3}}$ **e)** 27^0

6 a) $64^{\frac{1}{2}}$ **b)** $64^{\frac{-1}{3}}$ **c)** 64^0 **d)** $64^{\frac{2}{3}}$ **e)** $64^{\frac{5}{6}}$

7 a) $16^{\frac{1}{2}}$ **b)** $16^{\frac{-1}{4}}$ **c)** 16^0 **d)** $16^{\frac{3}{2}}$ **e)** $16^{\frac{7}{4}}$

8 a) $2^2 \times 9^{\frac{1}{2}}$ **b)** $2^5 \times 8^{\frac{1}{3}}$ **c)** $81^{\frac{1}{4}} \times 3^{-2}$ **d)** $9^{\frac{1}{2}} \times 6^2 \times 4^{-1}$

9 a) $25^{\frac{3}{2}}$ **b)** $36^{\frac{1}{2}}$ **c)** $125^{\frac{2}{3}} \times 8^{\frac{2}{3}}$ **d)** $49^{\frac{3}{2}} \times 81^{\frac{-1}{4}}$

10 a) $2^2 + 3^0 + 16^{\frac{1}{2}}$ **b)** $\left(\frac{3}{4}\right)^{-2} \times 27^{\frac{2}{3}}$ **c)** $4^2 \div 9^{\frac{1}{2}}$ **d)** $4^2 - 8^{\frac{1}{3}} + 9^0$

11 a) $5^{-2} \times 10^5 \times 16^{\frac{-1}{2}}$ **b)** $\left(\frac{4}{5}\right)^2 \times 128^{\frac{-3}{7}}$ **c)** $5^3 - 25^{\frac{1}{2}} - \left(\frac{2}{5}\right)^{-2}$ **d)** $125^{\frac{1}{3}} - 121^{\frac{1}{2}} + 216^{\frac{1}{3}}$

EXERCISE 4.1 continued

You may use a calculator for questions **12** to **20**.
Give the answers either exactly or correct to 5 significant figures.

12 a) $1\cdot14^5$ **b)** $2\cdot79^3$ **c)** $1\cdot005^9$ **d)** $4\cdot1^{-4}$

13 a) $3\cdot25^4$ **b)** $0\cdot46^5$ **c)** $1\cdot01^7$ **d)** $2\cdot91^{-3}$

14 a) $923\,521^{\frac{1}{4}}$ **b)** $1\cdot051^{\frac{1}{5}}$ **c)** $21^{\frac{1}{7}}$ **d)** $6\cdot45^{\frac{2}{5}}$

15 a) $14\,641^{\frac{1}{4}}$ **b)** $14\,120^{\frac{1}{5}}$ **c)** $9^{\frac{1}{9}}$ **d)** $1024^{\frac{2}{5}}$

16 a) $100 \times 1\cdot02^3$ **b)** $1\cdot6^5 \times 2\cdot1^{\frac{1}{3}}$ **c)** $(10^5 \times 4\cdot1)^{\frac{1}{4}}$

17 a) $4^3 + 3^4$ **b)** $1\cdot6^4 \times 1\cdot7^{\frac{1}{4}}$ **c)** $1^5 \times 4\cdot1^{\frac{1}{4}}$

18 a) $1\cdot9^4 - 2\cdot1^3$ **b)** $1\cdot9^{\frac{1}{4}} + 0\cdot97^{\frac{1}{5}}$ **c)** $14^3 - 196^{\frac{3}{2}}$

19 a) $5\cdot27^5 - 3\cdot49^5$ **b)** $4^{\frac{3}{4}} + 5^{\frac{2}{5}}$ **c)** $216^{\frac{4}{3}} \times 9^{\frac{-2}{3}}$

20 Work out these.
Give your answers either as whole numbers or as fractions.

   a) $3^2 \times 4^{\frac{1}{2}}$ **b)** $3^4 \times 9^{\frac{1}{2}}$ **c)** $125^{\frac{1}{3}} \times 5^{-2}$ **d)** $16^{\frac{1}{2}} \times 6^2 \times 2^{-3}$

   e) $2^3 + 4^0 + 49^{\frac{1}{2}}$ **f)** $\left(\frac{1}{2}\right)^{-3} \times 27^{\frac{2}{3}}$ **g)** $6^2 \div 25^{\frac{1}{2}}$ **h)** $5^2 + 8^{\frac{1}{3}} - 7^0$

C CHALLENGE 1

Here is a table for $y = 2^x$.

x	0	1	2	3	4
y = 2ˣ	1	2	4	8	16

a) Plot a graph of $y = 2^x$.

b) Use your graph to suggest values for $2^{\frac{1}{2}}$, $2^{\frac{1}{3}}$ and $2^{\frac{3}{2}}$.

Use the rule $(a^m)^n = a^{m \times n}$ to find the value of $\left(2^{\frac{1}{2}}\right)^2$.

What does this suggest that $2^{\frac{1}{2}}$ means?
Does this agree with the answer you read off the graph?

c) Similarly, use the rule $(a^m)^n = a^{m \times n}$ to find the value of $\left(2^{\frac{1}{3}}\right)^3$.

What does this suggest that $2^{\frac{1}{3}}$ means?
Does this agree with the value you read off the graph?

d) Again use the rule $(a^m)^n = a^{m \times n}$ to find the value of $\left(2^{\frac{3}{2}}\right)^2$.

What does this suggest that $2^{\frac{3}{2}}$ means?
Does this agree with the value you read off the graph?

Using the laws of indices with numbers and letters

You have already learned the laws for indices.

$$a^n \times a^m = a^{n+m}, \quad a^n \div a^m = a^{n-m}, \quad (a^n)^m = a^{n \times m}$$

These can be used with either numbers or letters.

EXAMPLE 4

Write each of these as a single power of 2, where possible.

a) $2\sqrt{2}$ **b)** $\left(\sqrt[3]{2}\right)^2$ **c)** $2^3 \div 2^{\frac{1}{2}}$ **d)** $2^3 + 2^4$

e) $8^{\frac{3}{4}}$ **f)** $2^3 \times 4^{\frac{3}{2}}$ **g)** $2^n \times 4^3$

a) $2\sqrt{2} = 2^1 \times 2^{\frac{1}{2}} = 2^{\frac{3}{2}}$

b) $\left(\sqrt[3]{2}\right)^2 = (2^{\frac{1}{3}})^2 = 2^{\frac{2}{3}}$

c) $2^3 \div 2^{\frac{1}{2}} = 2^{3-\frac{1}{2}} = 2^{2\frac{1}{2}} = 2^{\frac{5}{2}}$

d) $2^3 + 2^4$ These powers cannot be added.

e) $8^{\frac{3}{4}} = (2^3)^{\frac{3}{4}} = 2^{3 \times \frac{3}{4}} = 2^{\frac{9}{4}}$

f) $2^3 \times 4^{\frac{3}{2}} = 2^3 \times (2^2)^{\frac{3}{2}} = 2^3 \times 2^3 = 2^6$

g) $2^n \times 4^3 = 2^n \times (2^2)^3 = 2^n \times 2^6 = 2^{n+6}$

> **EXAM TIP**
> The most common mistake is trying to add or subtract a^x and a^y, which cannot be done.

EXAMPLE 5

Write 132 as a product of its prime factors.

Use indices where possible.

$$\begin{aligned} 132 &= 2 \times 66 \\ &= 2 \times 2 \times 33 \\ &= 2 \times 2 \times 3 \times 11 \\ &= 2^2 \times 3 \times 11 \end{aligned}$$

EXERCISE 4.2

1 Write each of these numbers as a power of 3, as simply as possible.

a) 27

b) $\frac{1}{3}$

c) $3 \times \sqrt{3}$

d) $81^{\frac{3}{2}}$

e) $3^4 \times 9^{-1}$

f) $9^n \times 27^{3n}$

2 Write each of these numbers as a power of 2, as simply as possible.

a) 32

b) $8^{\frac{2}{3}}$

c) $2 \times \sqrt[3]{64}$

d) $0 \cdot 25$

e) $2^{2n} \times 4^{\frac{n}{2}}$

f) $2^{3n} \times 16^{-2}$

3 Write each of these numbers as a power of 5, as simply as possible.

a) 625

b) $25^{\frac{-1}{2}} \times 5^3$

c) $0 \cdot 2$

d) $125^{\frac{3}{2}} \times 5^{-3} \div 25^2$

e) $5^4 - 5^3$

f) $25^{3n} \times 125^{\frac{n}{3}}$

4 Write each of these numbers as a power of a prime number, as simply as possible.

a) 343

b) $25^{\frac{1}{6}}$

c) $16^{\frac{1}{2}} \times 64^{\frac{-2}{3}}$

d) $27^2 \div 81^{\frac{3}{2}}$

e) $2^5 + 2^2$

f) $9^{2n} \times 3^{-2n}$

5 Write each of these numbers as powers of 2 and 3, as simply as possible.

a) 24

b) $6^2 \times 4^2$

c) $18^{\frac{1}{3}}$

d) $\frac{4}{9}$

e) $13\frac{1}{2}$

f) 12^{2n}

6 Write each of these numbers as a product of its prime factors.
Use indices where possible.

a) 36

b) 96

c) 60

d) 392

e) 75

f) 144

g) 300

h) 324

7 Write each of these numbers as a product of powers of prime numbers.

a) 15^3

b) $12^{\frac{1}{2}} \times 9^{\frac{-1}{4}}$

c) 40^n

d) $20^{2n} \times 100^n$

8 Write each of these numbers as a power of 5, as simply as possible. If any cannot be simplified, say so.

a) $0 \cdot 2$

b) 125

c) 25^2

d) $125 \times 5^{-2} \times 25^2$

e) $5^4 - 5^3$

f) $25^{3n} \times 125^{\frac{n}{3}}$

K KEY IDEAS

- The laws of indices are $a^n \times a^m = a^{m+n}$, $a^n \div a^m = a^{n-m}$, $(a^n)^m = a^{n \times m}$, $a^0 = 1$, $a^{-n} = \frac{1}{a^n}$, $a^{\frac{1}{n}} = \sqrt[n]{a}$

- You can work out powers using the $\boxed{\wedge}$ key on your calculator.

5 Rearranging formulae

All the formulae that have been covered previously contained the new subject only once, and always as part of the numerator. This is now extended.

EXAMPLE 1

Rearrange the formula $a = x + \dfrac{cx}{d}$ to make x the subject.

$$a = x + \frac{cx}{d}$$

$$ad = dx + cx \qquad \text{Multiply through by } d.$$

$$dx + cx = ad \qquad \text{Rearrange to get all the terms involving } x \text{ on the left-hand side.}$$

$$x(d + c) = ad \qquad \text{Factorise.}$$

$$x = \frac{ad}{d + c} \qquad \text{Divide by } (d + c).$$

EXAMPLE 2

Rearrange the formulae $a = \dfrac{1}{p} + \dfrac{1}{q}$ to make p the subject.

$$a = \frac{1}{p} + \frac{1}{q}$$

$apq = q + p$ — Multiply through by pq.

$apq - p = q$ — Collect all terms in p on the left-hand side.

$p(aq - 1) = q$ — Factorise.

$p = \dfrac{q}{aq - 1}$ — Divide by $(aq - 1)$.

EXAMPLE 3

Rearrange the formula $a = b + \dfrac{c}{1 + p}$ to make p the subject.

$$a = b + \frac{c}{1 + p}$$

$a(1 + p) = b(1 + p) + c$ — Multiply through by $(1 + p)$.

$a + ap = b + bp + c$ — Expand brackets.

$ap - bp = b + c - a$ — Collect all terms in p on the left-hand side.

$p(a - b) = b + c - a$ — Factorise.

$p = \dfrac{b + c - a}{a - b}$ — Divide by $(a - b)$.

EXAMPLE 4

Rearrange the formula $s = b + \sqrt{\dfrac{c}{p}}$ to make c the subject.

$$s = b + \sqrt{\dfrac{c}{p}}$$

$$s - b = \sqrt{\dfrac{c}{p}}$$ If a root or power is involved, rearrange to get that by itself.

$$(s - b)^2 = \dfrac{c}{p}$$ Square both sides.

$$\dfrac{c}{p} = (s - b)^2$$ Rearrange to get all the terms in c on the left-hand side.

$$c = p(s - b)^2$$ Multiply through by p.
In this case there is no need to expand $(s - b)^2$.

EXERCISE 5.1

Rearrange each of these equations to make the letter in square brackets the subject.

1 $s = at + 2bt$ [t]

2 $s = ab - bc$ [b]

3 $P = t - \dfrac{at}{b}$ [t]

4 $s = \dfrac{1}{a} + b$ [a]

5 $s - 2ax = b(x - s)$ [x]

6 $3(a + y) = by + 7$ [y]

7 $a = \dfrac{t}{b} - st$ [t]

8 $a = \dfrac{1}{b + c}$ [c]

9 $2(a - 1) = b(1 - 2a)$ [a]

10 $\dfrac{a}{b} - 2a = b$ [a]

11 $a = b + \dfrac{c}{d + 1}$ [d]

12 $m = \dfrac{100(a - b)}{b}$ [b]

13 $a = b + c^2$ [c]

14 $A = P + \dfrac{PRT}{100}$ [P]

15 $\dfrac{a}{p} = \dfrac{1}{1 + p}$ [p]

16 $a = \dfrac{1}{1 + x} - b$ [x]

17 $\dfrac{a}{x + 1} = \dfrac{b}{2x + 1}$ [x]

18 $T = 2\pi\sqrt{\dfrac{L}{g}}$ [L]

19 $s = 2r^2 - 1$ [r]

EXERCISE 5.1 continued

20 $s = \dfrac{uv}{u+v}$ $[v]$

21 $s = ut + \dfrac{at}{2}$ $[t]$

22 $\dfrac{a}{2x+1} = \dfrac{b}{3x-1}$ $[x]$

23 $\dfrac{1}{f} = \dfrac{1}{u} + \dfrac{1}{v}$ $[v]$

24 $y = 3x^2 - 4$ $[x]$

25 $A = \pi r \sqrt{h^2 + r^2}$ $[h]$

26 $v^2 - u^2 = 2as$ $[u]$

27 $V = \frac{1}{3}\pi r^2 h$ $[r]$

28 $s = 15 - \frac{1}{2}at^2$ $[t]$

K KEY IDEAS

- When rearranging a formula where the new subject occurs more than once, collect together all terms containing the new subject.

- When a formula involves a power or root, rearrange the formula to get that term by itself.

- When the new subject is raised to a power, use the inverse operation. For example, for $\sqrt{x}$, square; for x^2, take the square root.

STAGE
9

6 Arcs, sectors and volumes

You will learn about

- Finding the length of an arc and the area of a sector of a circle
- Finding the volume of cones, spheres and pyramids

You should already know

- How to find the circumference and area of a circle
- How to find the volume of a prism
- How to rearrange formulae
- How to use Pythagoras' theorem and trigonometry

Arcs and sectors

A ACTIVITY 1

a) How many of the sectors in the diagram on the right would fit into a circle of radius 5 cm?

b) Use your answer to work out the area of one of these sectors and the arc length of the sector.

c) How would you find the perimeter of the sector?

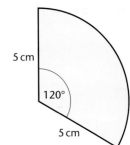

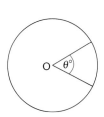

A sector is a fraction of a circle. It is $\dfrac{\theta}{360}$ of the circle, where $\theta°$ is the sector angle at the centre of the circle.

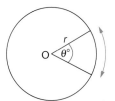

Arc length $= \dfrac{\theta}{360} \times$ circumference

$= \dfrac{\theta}{360} \times 2\pi r$

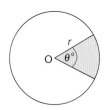

Sector area $= \dfrac{\theta}{360} \times$ area of circle

$= \dfrac{\theta}{360} \times \pi r^2$

EXAMPLE 1

Calculate the arc length and area of this sector.

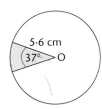

5·6 cm

37° O

Arc length $= \dfrac{\theta}{360} \times 2\pi r$

$= \dfrac{37}{360} \times 2\pi \times 5\!\cdot\!6$

$= 3\!\cdot\!62\,\text{cm}$ to 3 significant figures.

Sector area $= \dfrac{\theta}{360} \times \pi r^2$

$= \dfrac{37}{360} \times \pi \times 5\!\cdot\!6^2$

$= 10\!\cdot\!1\,\text{cm}^2$ to 3 significant figures.

STAGE
9

EXAMPLE 2

Calculate the sector angle of a sector with arc length 6·2 cm in a circle with radius 7·5 cm.

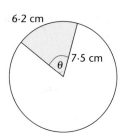
6·2 cm
7·5 cm
θ

Arc length $= \dfrac{\theta}{360} \times 2\pi r$

$6·2 = \dfrac{\theta}{360} \times 2\pi \times 7·5$

$\theta = \dfrac{6·2 \times 360}{2\pi \times 7·5}$

$= 47·4°$ to 3 significant figures.

EXAMPLE 3

A sector makes an angle of 54° at the centre of a circle. The area of the sector is 15 cm².

Calculate the radius of the circle.

Sector area $= \dfrac{\theta}{360} \times \pi r^2$

$15 = \dfrac{54}{360} \times \pi r^2$

$r^2 = \dfrac{15 \times 360}{54 \times \pi}$

$= 31·83...$

$r = \sqrt{31·83...}$

$= 5·64$ cm to 3 significant figures.

STAGE
9

EXAM TIP

You can rearrange the formula before you substitute, if you prefer.

EXERCISE 6.1

1 Calculate the arc length of each of these sectors.
Give your answers to 3 significant figures.

a)

b)

c)
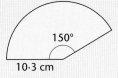

d)
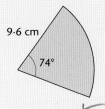

e)
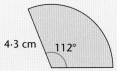

f)
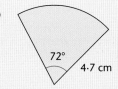

g)
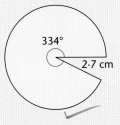

h)
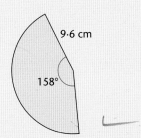

i)
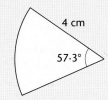

j)
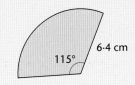

2 Calculate the area of each of the sectors in question **1**.
Give your answers to 3 significant figures.

STAGE
9

49

EXERCISE 6.1 continued

3 Calculate the perimeter of each of these sectors. Give your answers to 3 significant figures.

a)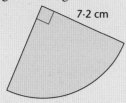
7·2 cm

b)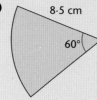
8·5 cm
60°

c)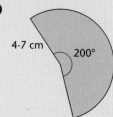
4·7 cm
200°

d)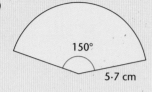
150°
5·7 cm

e)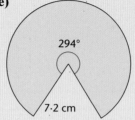
294°
7·2 cm

f)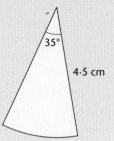
35°
4·5 cm

4 Calculate the sector angle in each of these sectors. Give your answers to the nearest degree.

a)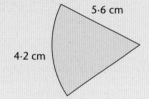
5·6 cm
4·2 cm

b)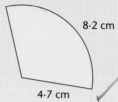
8·2 cm
4·7 cm

c)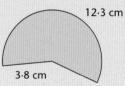
12·3 cm
3·8 cm

d)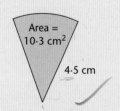
Area = 10·3 cm²
4·5 cm

e)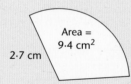
Area = 9·4 cm²
2·7 cm

f)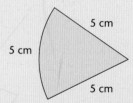
5 cm
5 cm
5 cm

g)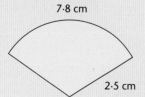
7·8 cm
2·5 cm

h)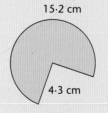
15·2 cm
4·3 cm

i)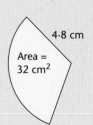
4·8 cm
Area = 32 cm²

j)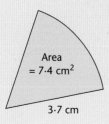
Area = 7·4 cm²
3·7 cm

k)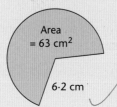
Area = 63 cm²
6·2 cm

l)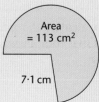
Area = 113 cm²
7·1 cm

5 Calculate the radius of each of these sectors.

a)

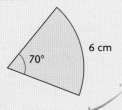

70° 6 cm

b)

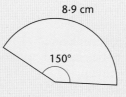

8·9 cm

150°

c)

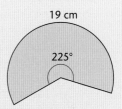

19 cm

225°

d)

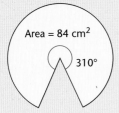

Area = 84 cm²

310°

e)

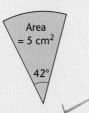

Area = 5 cm²

42°

f)

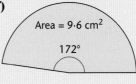

Area = 9·6 cm²

172°

g)

42°

9·8 cm

h)

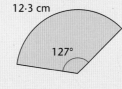

12·3 cm

127°

i)

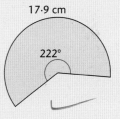

17·9 cm

222°

6 Calculate the radius of each of these sectors.
   a) Sector area = 19·7 cm², sector angle = 52°.
   b) Sector area = 2·7 cm², sector angle = 136°.
   c) Sector area = 6·2 m², sector angle = 218°.

7 A motif consists of two separate sectors of a circle, each with angle 35° and radius 32 mm. They are to be painted blue with a thin black border.

Calculate the total area of blue and the length of the black border required.
Give your answers to an appropriate degree of accuracy.

8 A lawn is in the shape of a sector of a circle with angle 63° and radius 25 m.
 a) The owner wants to spread fertiliser on the lawn. Calculate the area that needs to be covered.
 b) The owner wants to put edging around the lawn. Calculate the length of edging needed.
 Give your answers to an appropriate degree of accuracy.

9 A sector of a circle of radius 6·2 cm has an arc length of 20·2 cm. Calculate the angle of the sector and hence find the area of the sector.

10 A cushion is in the shape of a sector with a sector angle of 150°. The radius is 45 cm. Edging is sewn all round the cushion. What length of edging is required?

C CHALLENGE 1

A sector of a circle is joined to form a cone.

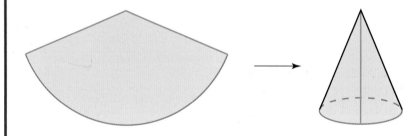

Find the radius of the base of the cone made with sectors of these dimensions.

a) Radius 5·8 cm, angle 162°

b) Radius 9·7 cm, angle 210°

c) Radius 12·1 cm, angle 295°

STAGE
9

Volumes

In Stage 7 you met the formula for the volume of a prism – a shape where the cross-section stays the same throughout its length.

Volume of a prism = area of cross-section × length

For a shape whose cross-section, though similar, decreases to a point, the volume is given by

Volume = $\frac{1}{3}$ × area of cross-section at base × height

Shapes that decrease to a point in this way include the pyramid, with a square or rectangular cross-section, and the cone. So, for a cone with base radius r and height h,

Volume = $\frac{1}{3}\pi r^2 h$

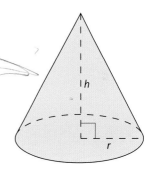

You also need to know about a different type of three-dimensional shape – the sphere. For a sphere of radius r,

Volume = $\frac{4}{3}\pi r^3$

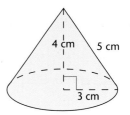

▌ EXAMPLE 4

Find the volume of this cone.

Volume = $\frac{1}{3}\pi r^2 h$

= $\frac{1}{3}\pi \times 3^2 \times 4$

= 12π

= $37 \cdot 7 \, \text{cm}^3$ to 3 significant figures.

STAGE
9

EXAMPLE 5

A bowl is in the shape of a hemisphere of diameter 25 cm.

Calculate its volume, in litres.

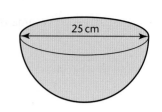
25 cm

Volume of hemisphere $= \frac{1}{2} \times$ volume of sphere

$$= \frac{1}{2} \times \frac{4}{3}\pi r^3$$

$$= \frac{2}{3}\pi r^3$$

$$= \frac{2}{3}\pi \times 12 \cdot 5^3$$

$$= 4090 \cdot 6 \ldots \text{cm}^3 \qquad\qquad 1 \text{ litre} = 1000 \text{ cm}^3$$

$$= 4 \cdot 09 \text{ litres to 3 significant figures.}$$

EXERCISE 6.2

1 Calculate the volume of each of these pyramids. Their bases are squares or rectangles.

a)

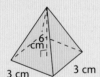
6 cm
3 cm   3 cm

b)

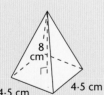
8 cm
4·5 cm   4·5 cm

c)

6 m
5 m   7 m

d)

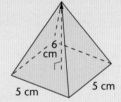
6 cm
5 cm   5 cm

e)

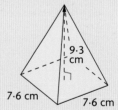
9·3 cm
7·6 cm   7·6 cm

f)

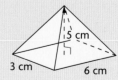
5 cm
3 cm   6 cm

2 Calculate the volume of each of these cones.

a)

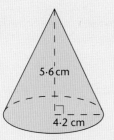

5·6 cm
4·2 cm

b)

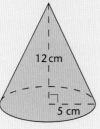

12 cm
5 cm

c)

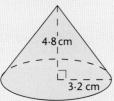

4·8 cm
3·2 cm

d)

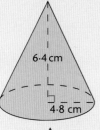

6·4 cm
4·8 cm

e)

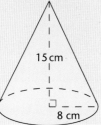

15 cm
8 cm

f)

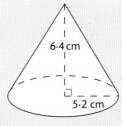

6·4 cm
5·2 cm

3 Find the volume of a sphere of these radii.
a) 5 cm
b) 6·2 cm
c) 2 mm
d) 3 cm
e) 4·7 cm
f) 7·8 mm

4 A cone has a base of radius 4·2 cm and a slant height of 7·8 cm.
a) Find its height.
b) Calculate its volume.

5 A pyramid has a square base with sides of 8 cm.
Its volume is 256 cm³.
Find its height.

6 Find the radius of the base of a cone with these dimensions.
a) Volume 114 cm³, height 8·2 cm
b) Volume 52·9 cm³, height 5·4 cm
c) Volume 500 cm³, height 12·5 cm

7 Find the volume of a sphere with these dimensions.
a) Radius 5·1 cm
b) Radius 8·2 cm
c) Diameter 20 cm

8 Find the radius of a sphere of these volumes.
a) 1200 cm³
b) 8000 cm³

9 Calculate the capacity of this glass. Give your answer in millilitres.

4 cm
7 cm

EXERCISE 6.2 continued

10 This glass paperweight in the shape of a cone has a volume of 75 cm³.
Its base radius is 3 cm.

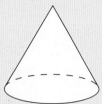

Calculate its height.

11 How many ball bearings of radius 0·3 cm can be made from 10 cm³ of metal when it is melted?

12 How many glass marbles of radius 7 mm can be made from 100 cm³ of glass?

13 A plastic pipe is a cylinder 2 m long. The internal and external radii of the pipe are 5 mm and 6 mm.

Calculate the volume of plastic in the pipe.

14 A solid cone and a solid cylinder both have base radius 6 cm.
The height of the cylinder is 4 cm.
The cone and the cylinder both have the same volume.
Find the height of the cone.

15 A sphere has the same volume as this cone.

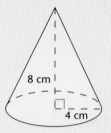

Calculate the radius of the sphere.

C CHALLENGE 2

A sphere of radius 12·8 cm has the same volume as a cone of base radius 8·0 cm.

Find the height of the cone.

KEY IDEAS

■ Arc length $= \dfrac{\theta}{360} \times 2\pi r$

■ Sector area $= \dfrac{\theta}{360} \times \pi r^2$

■ Volume of a prism = area of cross-section × length

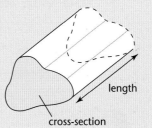

length

cross-section

■ Volume of a pyramid $= \dfrac{1}{3} \times$ length × width × height

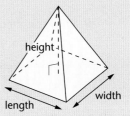

height

length   width

■ Volume of a cone $= \dfrac{1}{3}\pi r^2 h$

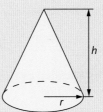

h

r

■ Volume of a sphere $= \dfrac{4}{3}\pi r^3$

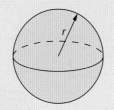

r

Arcs, sectors and volumes

6

STAGE
9

Upper and lower bounds

7

STAGE
9

You will learn about

- The upper and lower bounds of measurements and the effect of these on calculations

You should already know

- About degrees of accuracy; for example, that the value of '18·6 seconds to 1 decimal place' lies between 18·55 and 18·65 seconds

Sums and differences of measurements

You should already know that, if $t = 18·6$ s to 1 decimal place, then

$$18·55\,\text{s} \leqslant t < 18·65\,\text{s}$$

This is called the lower bound.

This is called the upper bound.

EXAM TIP

Many people are confused about the upper bound. The convention is that the lower bound is contained in the interval and the upper bound is in the next, higher interval. The measured value 18·6 could be as near 18·65 as you like – 18·6499 or 18·649 999 for instance, but the upper bound is 18·65, not even 18·649.

Consider two kitchen cupboards, of width 300 mm and 500 mm correct to the nearest millimetre.

	300 mm	500 mm
Lower bound of width:	299·5 mm	499·5 mm
Upper bound of width:	300·5 mm	500·5 mm

If the cupboards are put next to each other then

the lower bound of w, their joint width = 299·5 + 499·5 = 799 mm and
the upper bound of w, their joint width = 300·5 + 500·5 = 801 mm

so 799 mm $\leq w <$ 801 mm.

> **To find the lower bound of a sum, add the corresponding lower bounds.**
> **Similarly, to find the upper bound of a sum, add the upper bounds.**

However, the difference between the widths of the kitchen cupboards is largest when the smallest possible width of the smaller cupboard is subtracted from the largest possible width of the larger cupboard; the difference is least when the largest possible width of the smaller cupboard is subtracted from the smallest possible width of the larger cupboard. So

the upper bound of the difference in their widths = 500·5 – 299·5 = 201 mm and
the lower bound of the difference in their widths = 499·5 – 300·5 = 199 mm.

> **To find the upper bound of a difference, subtract the lower bound of the smaller value from the upper bound of the larger value.**

> **To find the lower bound of a difference, subtract the upper bound of the smaller value from the lower bound of the larger value.**

EXAMPLE 1

A piece of red ribbon is 35·2 cm to the nearest millimetre. A piece of blue ribbon is 12·6 cm to the nearest millimetre.

a) What is the minimum length of the two pieces of ribbon laid end to end?

b) What is the lower bound of the difference in the lengths of the two pieces?

For the red piece: LB = 35·15 cm   UB = 35·25 cm.
For the blue piece: LB = 12·55 cm   UB = 12·65 cm.

a) Minimum total length = sum of lower bounds
= 35·15 + 12·55
= 47·7 cm

b) Lower bound of difference in lengths = LB of longer piece – UB of shorter piece
= 35·15 – 12·65
= 22·5 cm

STAGE
9

59

1 Calculate the upper bound of the sum of each of these pairs of measurements.
   a) 29·7 seconds and 31·4 seconds (both to 3 significant figures)
   b) 11·04 seconds and 13·46 seconds (both to the nearest hundredth of a second)
   c) 6·42 m and 5·97 m (both to the nearest centimetre)
   d) 1·248 kg and 0·498 kg (both to the nearest gram)
   e) 86 mm and 98 mm (both to the nearest millimetre)
   f) 11·042 kg and 1·695 kg (both to the nearest gram)
   g) 78·5 cm and 69·7 cm (both to 3 significant figures)
   h) 46·03 s and 59·82 s (both to the nearest $\frac{1}{100}$ second)

2 Find the lower bound of the sum of each pair of measurements in question **1**.

3 Find the upper bound of the difference between each of these pairs of measurements.
   a) 947 g and 1650 g (to the nearest g)
   b) 16·4 cm and 9·8 cm (to the nearest mm)
   c) 1650 g and 870 g (to the nearest 10 g)
   d) 24·1 s and 19·8 s (to the nearest 0·1 s)
   e) 14·86 s and 15·01 s (to the nearest $\frac{1}{100}$ second)
   f) 493 m and 568 m (to the nearest m)
   g) 12 700 m and 3800 m (to the nearest 100 m)
   h) 1·824 g and 1·687 g (to the nearest mg)

4 Calculate the lower bound of the difference between each pair of measurements in question **3**.

5 A piece of paper 21·0 cm long is taped on to the end of another piece 29·7 cm long, both measurements given to the nearest millimetre. What is the upper bound of the total length?

6 Two stages of a relay race are run in times of 14·07 seconds and 15·12 seconds to the nearest 0·01 second.
   a) Calculate the upper bound of the total time for these two stages.
   b) Calculate the upper bound of the difference between the times for these two stages.

7 A triangle has sides of length 7 cm, 8 cm and 10 cm.
   All measurements are correct to the nearest centimetre.
   Work out the upper and lower bounds of the perimeter of the triangle.

8 Stuffing weighing 0·5 kg is added to a chicken weighing 2·4 kg.
   Both weights are correct to the nearest 0·1 kg.
   What are the maximum and minimum possible weights of the stuffed chicken?

9 Phil and Ruby each draw a line 15 cm long, to the nearest centimetre.
   What are the greatest and least possible differences between the lengths drawn?

10 Given that $p = 5·1$ and $q = 8·6$, both correct to 1 decimal place, work out the largest and smallest possible values of
   a) $p + q$.
   b) $q - p$.

Multiplying and dividing measurements

Consider a piece of A4 paper whose measurements are given as 21·0 cm and 29·7 cm to the nearest millimetre. What are the upper and lower bounds of the area of the piece of paper?

29·75 cm

Upper bound   21·05 cm

Upper bound of area
= 29·75 × 21·05
= 626·2375 cm^2

29·65 cm

Lower bound   20·95 cm

Lower bound of area
= 29·65 × 20·95
= 621·1675 cm^2

When multiplying
- **to find the upper bound, multiply the upper bounds.**
- **to find the lower bound, multiply the lower bounds.**

When dividing, however, the situation is different.

Dividing by a larger number makes the answer smaller.

When dividing
- **to find the upper bound, divide the upper bound by the lower bound.**
- **to find the lower bound, divide the lower bound by the upper bound.**

EXAMPLE 2

Pete cycles 14·2 km (to 3 significant figures) in a time of 46 minutes (to the nearest minute).

What is the upper bound of his average speed in kilometres per hour?
Give your answer to 3 significant figures.

$$\text{Upper bound of speed} = \frac{\text{upper bound of distance}}{\text{lower bound of time}}$$

$$= \frac{14\cdot25}{45\cdot5} \text{ km/minute}$$

$$= \frac{14\cdot25}{45\cdot5} \times 60 \text{ km/h}$$

$$= 18\cdot8 \text{ km/h to 3 significant figures}$$

EXAM TIP
To find the upper bound of any combined measurement, work out which of the upper and lower bounds of the given measurements you need to use to give you the greatest result. If you aren't sure, experiment!

STAGE
9

EXERCISE 7.2

1 One make of car has a length of 3·2 m, to the nearest 0·1 m. A car transporter of length 16·1 m, to the nearest 0·1 m, needs to carry 5 cars. Will the 5 cars always fit on to the transporter?

2 Pencils have a width of 8 mm, to the nearest millimetre.
What is the smallest width of a pencil box that can hold 10 pencils side by side?

3 Find the upper bound of the floor area of a rectangular room with these measurements.
   a) 3·8 m by 4·2 m to 2 significant figures
   b) 5·26 m by 3·89 m to the nearest centimetre
   c) 8·42 m by 6·75 m to 3 significant figures
   d) 7·6 m by 5·2 m to the nearest 10 cm

4 Find the lower bound of the floor area of each room in question **3**.

5 Calculate the upper bound of the distance travelled for each of these pairs of speed and time.
   a) 92·4 cm/second for 12·3 seconds
   b) 1·54 m/second for 8·2 seconds
   c) 57 km/h for 2·5 hours
   d) 5·61 m/second for 2·08 seconds

6 Calculate the lower bound of each distance for the data in question **5**.

7 Find the lower bound of the average speed for these measured times and distances. Give your answers to 3 significant figures.
   a) 6·4 cm in 1·2 seconds
   b) 12·4 m in 9·8 seconds
   c) 106 m in 10·0 seconds

8 Calculate the upper bound of each speed for the measurements in question **7**, giving your answers to 3 significant figures.

9 The mass of an object is given as 1·657 kg to the nearest gram. Its volume is 72·5 cm³ to 3 significant figures. Find the upper bound of its density. Give your answer to 4 significant figures.

10 Calculate, to 3 significant figures, the minimum width of a rectangle with these dimensions.
   a) Area = 210 cm² to 3 significant figures,
   length = 17·89 cm to the nearest millimetre
   b) Area = 210 cm² to 2 significant figures,
   length = 19·2 cm to the nearest millimetre
   c) Area = 615 cm² to 3 significant figures,
   length = 30·0 cm to the nearest millimetre

11 Calculate, to 3 significant figures, the upper bound of the height of a cuboid with these dimensions.
   a) Volume = 72 cm³,
   length = 6·2 cm,
   width = 4·7 cm
   (all to 2 significant figures)
   b) Volume = 985 cm³,
   length = 17·0 cm,
   width = 11·3 cm
   (all to 3 significant figures)
   c) Volume = 84 m³,
   length = 6·2 m,
   width = 3·8 m
   (all to 2 significant figures)

12 The population of a town is 108 000 to the nearest 1000. Its area is given as 129 square miles.
Calculate the upper and lower bounds of its population density, giving your answers to 3 significant figures.

13 Work out the largest and smallest possible areas of a rectangle measuring 27 cm by 19 cm, where both lengths are correct to the nearest centimetre.

EXERCISE 7.2 continued

14   Jasbinder runs 100 m in 12·8 seconds.
The distance is correct to the nearest
metre and the time is correct to the
nearest 0·1 second.
Work out the upper and lower bounds
of Jasbinder's speed.

15   A bar of gold is a cuboid and
measures 8·6 cm by 4·1 cm by 2·4 cm.
All the measurements are correct to
the nearest millimetre.
   a)   Work out the largest and smallest
possible volumes of the bar of gold.
   b)   The density of gold is 19·3 g/cm^3,
correct to the nearest 0·1 g/cm^3.
Work out the greatest and least
possible masses of the gold bar.

16   Use the formula $P = \dfrac{V^2}{R}$ to work out
the upper and lower bounds of P when
$V = 6$ and $R = 1$ and both values are
correct to the nearest whole number.

17   Work out the upper and lower bounds
of $\dfrac{8\cdot1 - 3\cdot6}{11\cdot4}$.
Each value in the calculation is correct
to 1 decimal place.

KEY IDEAS

■ To find the upper bound of any combined measurement, work out which of the upper
and lower bounds of the given measurements you need to use to give you the greatest
result. To find the lower bound, you need the smallest result.

Here is a summary of how this works in practice.

■ To find the lower bound of a sum, add the corresponding lower bounds. Similarly, to
find the upper bound of a sum, add the upper bounds.

■ To find the upper bound of a difference, subtract the lower bound of the smaller value
from the upper bound of the larger value.

■ To find the lower bound of a difference, subtract the upper bound of the smaller value
from the lower bound of the larger value.

■ When multiplying
to find the upper bound, multiply the upper bounds.
to find the lower bound, multiply the lower bounds.

■ When dividing
to find the upper bound, divide the upper bound by the lower bound.
to find the lower bound, divide the lower bound by the upper bound.

STAGE
9

Revision exercise B1

1 Write these in index form as simply as possible.
 a) The reciprocal of n
 b) The cube root of m
 c) The square root of $\frac{1}{n}$

2 Write each of these as a whole number or a fraction.
 a) 4^{-1}
 b) 5^0
 c) $25^{\frac{1}{2}}$
 d) 2^4
 e) $8^{\frac{2}{3}}$
 f) $125^{-\frac{2}{3}}$
 g) $\left(\frac{1}{8}\right)^{-\frac{1}{3}}$
 h) $64^{\frac{5}{6}}$
 i) $\left(\frac{4}{9}\right)^{-\frac{1}{2}}$
 j) $\left(\frac{1}{12}\right)^{-2}$

3 Write each of these as a whole number or a fraction.
 a) $8^0 \times 25^2$
 b) $4^2 \times 25^{\frac{1}{2}}$
 c) $12^2 \times 4^{-2}$
 d) $6^3 \div 9^{\frac{3}{2}}$
 e) $5^2 - 4^3 + 3^4$
 f) $25^{\frac{3}{2}} \times 64^{\frac{1}{3}}$
 g) $14^2 \times 49^{-1}$
 h) $\left(\frac{4}{5}\right)^{-2} \times \left(\frac{16}{9}\right)^{\frac{1}{2}}$

4 Work out these. Give the answers either exactly or to 5 significant figures.
 a) $1\cdot43^3$
 b) $0\cdot87^5$
 c) 2^{12}
 d) $7\cdot9^{-4}$

5 Work out these. Give the answers either exactly or to 5 significant figures.
 a) $59049^{\frac{1}{5}}$
 b) $7\cdot9^{\frac{1}{4}}$
 c) $4000^{\frac{1}{6}}$
 d) $32768^{\frac{3}{5}}$

6 Write each of these as a power of a prime number, as simply as possible.
 a) 128
 b) 27^2
 c) $49^{\frac{1}{3}}$
 d) $9^2 \div 81^{-\frac{1}{2}}$
 e) $2^n \times 32^{n+1}$

7 Write each of these as a product of the prime factors. Use indices where possible.
 a) 40
 b) 90
 c) 136
 d) 588

8 Write each of these as a product of powers of prime numbers, as simply as possible.
 a) 15^2
 b) $40^{\frac{1}{3}}$
 c) $14^3 \times 56^{\frac{1}{3}}$
 d) $72^{\frac{3}{2}} \times 24^{-\frac{1}{2}}$

9 Rearrange each of these formulae to make a the subject.
 a) $3a + 5c = 2b - 5a$
 b) $3(a + 2b) = 2b + 5a$

10 Rearrange each of these formulae to make p the subject.
 a) $p + q = 2(q - 3p)$
 b) $t = \dfrac{2(p - 1)}{p}$
 c) $\dfrac{1}{p} = \dfrac{1}{q} + \dfrac{1}{s}$
 d) $T = p + \dfrac{2p}{q}$
 e) $\dfrac{1}{2p - 1} = \dfrac{2a}{p + 1}$
 f) $p^2 + 4a = 2b$

11 A sector has an angle of $75°$ and a radius of $6\cdot5\,\text{cm}$. Calculate
 a) the arc length.
 b) the sector area.

12 Calculate the sector angle of each of these sectors.
 a)

7·2 cm
8·3 cm
 b)

Area = 29 cm²
4·8 cm

13 Calculate the radius of the circle with each of these sectors.

a)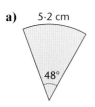
5·2 cm
48°

b)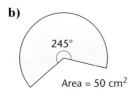
245°
Area = 50 cm²

14 Jo blows up a spherical balloon until it has a radius of 12 cm.
Find the volume of air she has blown into the balloon.

15 A concrete water tower has its internal volume in the shape of an inverted cone. The radius of the top is 3·6 m. The depth of the cone is 10 m.

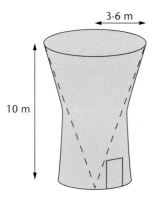
3·6 m
10 m

Calculate the volume of water which can be stored in the tower.

16 Paul measures out 250 g flour, 150 g butter and 120 g sugar, all to the nearest 10 g.
Calculate the upper and lower bounds of the total mass of flour, butter and sugar.

17 Two pieces of string measure 19·7 cm and 11·4 cm, to the nearest millimetre.
Calculate
a) the upper bound of the total length of the two pieces placed end to end.
b) the lower bound of the difference between the lengths of the two pieces.

18 The space for some kitchen base units is measured as 1000 mm to the nearest millimetre. Two base units are 500 mm each, to the nearest millimetre.
a) Explain why the two units will not necessarily fit into the space.
b) Calculate the upper bound of the gap remaining if the two units do fit in.

19 The length of a side of a cube is given as 4·6 cm to the nearest millimetre.
Calculate the upper and lower bounds of the volume of the cube, giving your answers to 3 significant figures.

20 A 100 m race was won in a time of 13·62 seconds, correct to the nearest hundredth of a second.
Calculate, to 3 significant figures, the upper bound of the average speed
a) if the distance is 100 m to 3 significant figures.
b) if the distance is 100·0 m to 4 significant figures.

21 Jane walks on an exercise machine for 7·2 minutes at a speed of 130 metres per minute, both measurements to 2 significant figures.
Calculate the upper bound of the distance she walks.

22 A town has a population of 94 300 to the nearest 100. Its area is 156 km², to the nearest square kilometre.
Calculate the lower bound of its population density, giving your answer to 3 significant figures.

STAGE
9

65

Similarity and enlargement

You will learn about

- Finding the volume and surface area of similar figures
- Enlargement using a negative scale factor

You should already know

- How to find volumes and areas
- How to use scale factors

The volumes and surface areas of similar figures

This cube has a volume of $8\,\text{cm}^3$.

This cube has a volume of $512\,\text{cm}^3$.

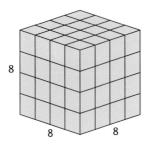

The lengths of the small cube have been enlarged with scale factor 4.

The volume has been enlarged with scale factor 64.

Since there are three dimensions for volume, and each dimension has been enlarged with scale factor 4, the volume scale factor = 4^3.

Similarly, consider the area of the face of each cube.

For the small cube, the area is $4\,cm^2$.

For the large cube, the area is $64\,cm^2$.

The area has been enlarged with scale factor 16.

There are two dimensions for area, so the area scale factor is 4^2.

For mathematically similar shapes
- **area scale factor = (length scale factor)^2**
- **volume scale factor = (length scale factor)^3.**

EXAMPLE 1

A model aircraft is made to a scale of $1:50$.

The area of the wing on the model is $18\,cm^2$.

What is the area of the wing on the real aircraft?

Length scale factor = 50

Area scale factor = 50^2

Area of real wing = 18×50^2
$= 45\,000\,cm^2$
$= 4\cdot5\,m^2$

There is an alternative solution to this problem.

With this scale, $2\,cm$ represent $1\,m$.

So $4\,cm^2$ represent $1\,m^2$.

So $18\,cm^2$ represent $\dfrac{18}{4} = 4\cdot5\,m^2$.

EXAM TIP
Remember that
$1\,m = 100\,cm$
$1\,m^2 = 10\,000\,cm^2$
$1\,m^3 = 1\,000\,000\,cm^3$

STAGE
9

EXAMPLE 2

A jug holding 50 cl is 12 cm high.

A similar jug holds 2 litres.

What is its height?

$$50\,cl = 0.5 \text{ litres}$$

$$\text{Volume scale factor} = \frac{2}{0.5} = 4$$

$$\text{Length scale factor} = \sqrt[3]{4}$$

$$\text{Height of larger jug} = 12\,cm \times \sqrt[3]{4}$$

$$= 19.0\,cm \text{ to 1 decimal place}$$

EXERCISE 8.1

1 State the area scale factor for each of these length scale factors.
  a) 2   **b)** 3
  c) 5   **d)** 4
  e) 6   **f)** 10

2 State the volume scale factor for each of these length scale factors.
  a) 10   **b)** 4
  c) 5   **d)** 2
  e) 3   **f)** 8

3 State the length scale factor for each of these.
  a) Area scale factor of 16
  b) Volume scale factor of 216
  c) Area scale factor of 64
  d) Volume scale factor of 1000

4 The model of a theatre set is made to a scale of 1:20.
  What area on the model will represent 1 m² on the real set?

5 A model of a building is made to a scale of 1:50.
  A room in the model is a cuboid with dimensions 7·4 cm by 9·8 cm by 6·5 cm high.
  Calculate the floor area of the room in
  a) the model.
  b) the actual building.

6 A medicine bottle holds 125 ml.
  How much does a similar bottle twice as high hold?

7 A glass holds 15 cl.
  The heights of this and a larger similar glass are in the ratio 1:1·2.
  Calculate the capacity of the larger glass.

8 A tray has an area of 160 cm².
  What is the area of a similar tray whose lengths are one and a half times as large?

9 The area of a table is 1·3 m².
  Calculate the area of a similar table whose sides are twice as long.

10 To what scale is a model drawn if an area of $5\,m^2$ in real life is represented by $20\,cm^2$ on the model?

11 The volumes of two similar jugs are in the ratio $1:4$.
What is the ratio of their heights?

12 Two mugs are similar.
One contains twice as much as the other.
The smaller one is $10\,cm$ high.
What is the height of the larger one?

13 $9\,m^2$ of fabric are required to cover a small sofa.
What area of fabric is required to cover a similar sofa $1\cdot2$ times as long?

14 Three similar wooden boxes have heights in the ratio $3:4:5$.
What is the ratio of their volumes?

15 Two vases are similar. The capacity of the smaller one is $250\,ml$. The capacity of the larger one is $750\,ml$. The height of the larger one is $18\,cm$.
Calculate the height of the smaller one.

16 A model of a room is made to a scale of $1:25$.
   a) What is the real height of a cupboard which is $8\,cm$ high on the model?
   b) What is the real area of a rug which has an area of $48\,cm^2$ on the model?
   c) What is the volume of a waste paper basket which has a volume of $1.2\,cm^3$ on the model?

17 The area of a rug is $2\cdot4$ times as large as the area of a similar rug.
The length of the smaller rug is $1\cdot6\,m$.
Find the length of the larger rug.

18 A model car is built to a scale of $1:24$.
It is $15\,cm$ long.
   a) How long is the real car?
   b) $10\,ml$ of paint are required to paint the model.
      How many litres of paint will be needed for the real car?

19 An artist makes a model for a large sculpture.
It is $24\,cm$ high.
The finished sculpture will be $3.6\,m$ high.
   a) Find the length scale factor.
   b) Find the area scale factor.
   c) The model has a volume of $1340\,cm^3$.
      What is the volume of the sculpture?

20 A wine glass is $12\,cm$ high.
How tall is a similar glass that holds twice as much?

21 A model aircraft is made to a scale of $1:48$.
The area of the wing of the real aircraft is $52\,m^2$.
What is the area of the wing of the model?

22 Barrels to hold liquid come in various sizes with different names.

An ordinary barrel holds 36 gallons.
A firkin holds 9 gallons.
A hogshead holds 54 gallons.
All the barrels are similar.

Find the ratio of their heights.

23 A square is enlarged by increasing the length of its sides by 10%.
The length of the sides was originally $8\,cm$.
What is the area of the enlarged square?

STAGE
9

C CHALLENGE 1

a) A simple model for the heat lost by birds is that it is proportional to their surface area.

Also, the energy a bird can produce, to replace the heat loss, is proportional to its volume.

Assuming there is a species of bird that occurs in two places with differing climate conditions, would you expect the birds in the colder place to be larger or smaller than those in the warmer place?

b) Investigate the relationship for similar aircraft between wing area (which gives the lift) and volume (which is closely associated with the mass).

Enlargement

To carry out an enlargement you need two pieces of information

- the scale factor
- the centre of enlargement.

If the scale factor of an enlargement is negative, the image is on the opposite side of the centre of enlargement from the object and the image is inverted. This is shown in the next example.

EXAMPLE 3

Plot the coordinates A(2, 2), B(4, 2) and C(2, 4) and join them to form a triangle.

Enlarge triangle ABC by a scale factor of ⁻3 using the point O(1, 1) as the centre of enlargement.

Plot the triangle ABC.

Then draw a line from each of the vertices through the centre of enlargement O and extend it on the other side.

Measure the distance from O to the vertex A and multiply it by 3.

Then mark the point A′ at a distance of 3 × OA along OA extended on the other side of the centre of enlargement.

Mark the points B′ and C′ in a similar way.

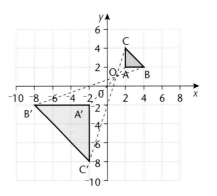

EXAM TIP

You can find the position of the image points by counting squares from the centre of enlargement as you have done previously so long as you remember that the image will be on the opposite side from the object. For example, from the centre of enlargement to point B is 3 units in the positive x direction and 1 unit in the positive y direction. Multiplying by the scale factor, point B′ is 9 units in the negative x direction and 3 units in the negative y direction.

If you are given both the original shape and the enlarged shape you can find the centre of enlargement as you have done previously. Join corresponding points on the two shapes with straight lines. The centre of enlargement is where the lines cross.

EXERCISE 8.2

1 Copy the diagram and find
  a) the centre of enlargement.
  b) the scale factor.

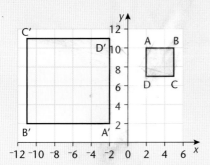

2 Draw a set of axes with the x-axis from 0 to 16 and the y-axis from 0 to 12.
Plot the points A(2, 8), B(6, 8), C(6, 4) and D(2, 4) and join them to form a square.
Enlarge the square by a scale factor of ⁻2 using the point O(5, 6) as the centre of enlargement.

STAGE
9

3 Copy the diagram and find
  a) the centre of enlargement.
  b) the scale factor.

4 Copy the diagram and find
  a) the centre of enlargement.
  b) the scale factor.

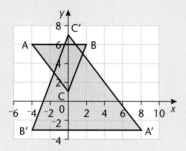

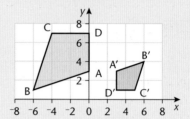

C CHALLENGE 2

a) Draw a scalene triangle.
Choose a centre and enlarge the triangle by a scale factor of ⁻1.
Can the same result be achieved with other transformations?

b) Can you find other transformations that give the same result if the triangle has reflection symmetry?

K KEY IDEAS

- For mathematically similar shapes
  - area scale factor = (length scale factor)^2
  - volume scale factor = (length scale factor)^3.

- If the scale factor of an enlargment is negative, the image is on the opposite side of the centre of enlargement from the object and the image is inverted.

Probability

You will learn about

- The addition rule for mutually exclusive events, P(A or B) = P(A) + P(B)
- The multiplication rule for independent events, P(A and B) = P(A) × P(B)
- Dependent events
- Conditional probability

You should already know

- How to draw and use a tree diagram
- The meaning of the terms *independent* and *mutually exclusive*

The addition rule for mutually exclusive events

In Stage 8, you learned that, if two events are **mutually exclusive**, they cannot occur at the same time and that the probability of either event A or event B happening can be found by adding the probabilities.

> If events A and B are mutually exclusive then P(A or B) = P(A) + P(B).

This can be extended to more than two events so long as all the events are mutually exclusive.

> If events A, B, C, … are mutually exclusive then
>
> P(A or B or C or …) = P(A) + P(B) + P(C) + … .

EXAMPLE 1

A card is taken at random from an ordinary pack of cards.

What is the probability that it is

a) a heart or a club?

b) a jack, a queen or a king?

c) a heart or an ace?

a) The events are mutually exclusive so the probabilities are added.

$$P(\text{heart or club}) = P(\text{heart}) + P(\text{club})$$
$$= \tfrac{1}{4} + \tfrac{1}{4}$$
$$= \tfrac{1}{2}$$

b) Again, the events are mutually exclusive so the probabilities are added.

$$P(\text{jack or queen or king}) = P(\text{jack}) + P(\text{queen}) + P(\text{king})$$
$$= \tfrac{1}{13} + \tfrac{1}{13} + \tfrac{1}{13}$$
$$= \tfrac{3}{13}$$

c) This time, as the ace of hearts is both a heart and an ace, the events are not mutually exclusive so you cannot simply add the probabilities.

To work the probability out you need to identify exactly which cards satisfy the requirements.

You can choose any of the 13 hearts (which include the ace of hearts), the ace of diamonds, the ace of spades or the ace of clubs.

So 16 of the 52 possible outcomes satisfy the requirements.

$$P(\text{heart or ace}) = \tfrac{16}{52}$$
$$= \tfrac{4}{13}$$

Alternatively, you can work out the probability by recognising that the ace of hearts is counted twice and taking this into account.

$$P(\text{heart or ace}) = P(\text{heart}) + P(\text{ace}) - P(\text{ace of hearts})$$
$$= \tfrac{13}{52} + \tfrac{4}{52} - \tfrac{1}{52}$$
$$= \tfrac{16}{52}$$
$$= \tfrac{4}{13}$$

Do not write the fractions in their simplest form as you need them to have a common denominator to add them.

The multiplication rule for independent events

When event A is unaffected by what has happened in another event, B, the events are **independent**. To find the probability of both event A and event B happening, the probabilities are multiplied.

In Stage 8, you learned how to use tree diagrams to deal with independent events when the outcomes were not equally likely. The examples which follow show how you can use the multiplication rule without drawing a tree diagram.

EXAMPLE 2

In a game at a school fair you have to first spin a coin and then roll an ordinary dice.

A prize is awarded if the coin shows heads and the score on the dice is less than 3.

Find the probability of winning a prize.

The probabilities are independent since the result on the coin does not affect the result on the dice.

The required combination of events is a head on the coin and a number less than 3 on the dice. As the events are independent, the probabilities are multiplied.

$P(\text{head}) = \frac{1}{2}$

$P(\text{number less than 3}) = P(1 \text{ or } 2)$

$$= \frac{2}{6}$$

$$= \frac{1}{3}$$

$P(\text{head and a number less than 3}) = \frac{1}{2} \times \frac{1}{3}$

$$= \frac{1}{6}$$

EXAMPLE 3

A jar contains 7 red balls and 4 white balls.

One ball is taken out and the colour noted. Then it is replaced before a second ball is removed.

What is the probability that

a) both balls are red?

b) the second ball is white?

The events are independent since the colour of the second ball is not affected by the colour of the first ball.

a) $P(\text{red}) = \frac{7}{11}$

Therefore $P(\text{red ball followed by red ball}) = \frac{7}{11} \times \frac{7}{11}$

$$= \frac{49}{121}$$

b) The second ball can be white when the first ball is red or when it is white.

On a tree diagram you would have followed two routes and added the results.

You use the same principles when you do not use a tree diagram.

You find the probabilities of the different ways of getting the desired combination and add them together.

$P(\text{white followed by white}) = \frac{4}{11} \times \frac{4}{11}$

$$= \frac{16}{121}$$

$P(\text{red followed by white}) = \frac{7}{11} \times \frac{4}{11}$

$$= \frac{28}{121}$$

Therefore $P(\text{second ball being white}) = \frac{16}{121} + \frac{28}{121}$

$$= \frac{44}{121}$$

$$= \frac{4}{11}$$

Notice that there is a shorter way to answer this.

Since the first ball can be either red or white, the probability is 1.

The probability of the second ball being white is $\frac{4}{11}$, so

$P(\text{second ball being white}) = 1 \times \frac{4}{11}$

$$= \frac{4}{11}$$

EXERCISE 9.1

1   There are 4 red counters, 5 white counters and 1 blue counter in a bag. If a counter is chosen at random, find the probability that it is red or blue.

2   Eileen is choosing her next car. The probability that she chooses a Ford is 0·5, the probability that she chooses a Rover is 0·35 and the probability that she chooses a Vauxhall is 0·15. Find the probability that Eileen chooses either a Ford or a Vauxhall.

3   There are 4 aces and 4 kings in a pack of 52 playing cards. I choose a card at random from the pack. What is the probability that it is an ace or a king?

4   There are 4 red counters, 5 white counters and 1 blue counter in a bag. I choose a counter at random, note its colour and put it back in the bag. I then do this a second time. Find the probability that both my choices are red.

5   The probability that I choose salad for school dinner is 0·6. The probability that I choose pizza for school dinner is 0·4. Assuming that the events are independent, what is the probability that I choose salad on Monday and pizza on Tuesday for school dinner?

6   What is the probability that I get a multiple of 2 when I throw a single fair dice? If I throw the dice twice, what is the probability that both throws give me a multiple of 2?

7   There are 12 picture cards in a pack of 52 playing cards. Geri selects a card at random, returns it to the pack and then randomly selects another card. Find the probability that both of Geri's selections are picture cards.

8   Emma and Rebecca are choosing where to go. The probability that Emma chooses to go to the cinema is 0·7. The probability that Rebecca chooses to go to the cinema is 0·8. Assuming that their choices are independent, find the probability that they both choose to go to the cinema.

9   The probability that the school football team will win their next game is 0·65. The probability that they will draw the next game is 0·2. What is the probability that they will win or draw their next game?

10   Assuming that the results of the football team are independent, use the probabilities in question 9 to find the probability that they will draw both of their next two games.

11   On Saturday morning Beverly either watches TV, plays on her computer or goes to her friend's house. She says that the probability that she watches TV is 0·4, the probability that she plays on her computer is 0·25 and the probability that she goes to her friend's house is 0·3. Why is her statement incorrect?

12 Andy is selecting a main course and a pudding from this menu.
The numbers next to the items are the probabilities that Andy chooses these items.

MENU

Main course	Pudding
Pizza (0·45)	Ice Cream (0·7)
Burger (0·3)	Fruit (0·3)
Fish Fingers (0·25)	

a) Find the probability that Andy chooses Pizza or Burger for his main course.

b) Assuming his choices are independent, find the probability that Andy chooses Fish Fingers and Ice Cream.

13 There are 4 aces in a pack of 52 playing cards.
I pick a card at random, replace it and then pick another card at random.

a) Find the probability, as a fraction in its simplest form, that the first card chosen is an ace.

b) Find the probability that both cards are aces.

14 The weather forecast says that there is a 60% chance that it will rain today and a 40% chance that it will rain tomorrow.

a) Write 60% and 40% as decimals.

b) Find the probability that it will rain on both days.

15 There is an equal likelihood that someone is born in any month of the year.
What is the probability that two people are both born in January?

16 I spin this 3-sided spinner three times. What is the probability that all my spins land on 2?

17 A bag contains 10 red balls, 5 blue balls and 8 green balls.
What is the probability of selecting

a) a red ball or a blue ball?

b) a green ball or a red ball?

18 When Mrs Smith goes to town the probability that she goes by bus is 0·5, the probability that she goes in a taxi is 0·35 and the probability that she goes on foot is 0·15.
What is the probability that she goes to town

a) by bus or by taxi?

b) by bus or on foot?

19 In any batch of computers made by a company, the probabilities of the number of faults per computer are as follows.

Number of faults	Probability
0	0·44
1	0·39
2	0·14
3	0·02
4	0·008
5	0·002

What is the probability that any particular computer will have

a) 1 or 2 faults?

b) 2, 3 or 4 faults?

c) an odd number of faults?

d) fewer than 2 faults?

e) at least 1 fault?

Probability

STAGE
9

EXERCISE 9.1 continued

20 The Channel 10 weather report says

> The probability of rain on Saturday is $\frac{3}{5}$ and the probability of rain on Sunday is $\frac{1}{2}$.
>
> This means it is certain to rain on Saturday or Sunday.

Explain why the report is wrong. What mistake have they made?

21 A coin is tossed and a dice is thrown. What is the probability of getting a head on the coin and an odd number on the dice?

22 This is the answer sheet to the multiple-choice section of a pub quiz. Bill's team are not very good and decide to randomly guess the answers. What is the probability that they guess all five answers correctly?

Answer sheet			
1 A	B	C	
2 A	B		
3 A	B	C	
4 A	B	C	D
5 A	B		

Conditional probability

In the previous section, you were working with independent events.

There are many situations, however, where the outcome of the second event is affected by the outcome of the first event. In this situation the probability of the second event depends on what has happened in the first event.

These events are not independent and are thus called **dependent events**.

In this situation the probability of the second event is called a **conditional probability** since it is conditional on the outcome of the first event.

In some situations you can work out the conditional probability yourself. In others you will be told what it is. These situations are illustrated in the following examples.

EXAMPLE 4

There are 4 red balls and 6 black balls in a bag.

If the first ball selected is black and is not replaced, what is the probability that the second ball is also black?

There are now 4 red and 5 black balls left in the bag so the probability is $\frac{5}{9}$.

EXAMPLE 5

There are 7 blue balls and 3 red balls in a bag.

A ball is selected at random and not replaced.

A second ball is then selected.

Find the probability that

a) two blue balls are chosen.

b) two balls of the same colour are chosen.

A tree diagram is a useful way to organise the work and the first step is the same as in the replacement situation.

For the second ball, however, the situation is different.

If the first ball was blue there are now 6 blue balls and 3 red balls left in the bag.

The probabilities are therefore $\frac{6}{9}$ and $\frac{3}{9}$.

If the first ball was red there are now 7 blue and 2 red balls left in the bag.

The probabilities are therefore $\frac{7}{9}$ and $\frac{2}{9}$.

The tree diagram looks like this.

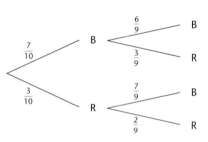

a) The probability that both balls are blue $= \frac{7}{10} \times \frac{6}{9}$

$$= \frac{42}{90}$$

$$= \frac{7}{15}$$

b) The probability that both balls are the same colour (i.e. both blue *or* both red)

$$= \frac{7}{10} \times \frac{6}{9} + \frac{3}{10} \times \frac{2}{9}$$

$$= \frac{42}{90} + \frac{6}{90}$$

$$= \frac{48}{90}$$

$$= \frac{8}{15}$$

You use the tree diagram in the same way as you would for independent events. The multiplication and addition rules still apply. It is the probabilities that are different from the independent case.

EXAM TIP

Although final answers should always be 'cancelled down' to their simplest form, it is usually unwise to cancel down the probabilities of the second (and third) event. This is because you often need to add the probabilities at the end and so you need them with a common denominator.

EXAMPLE 6

I have sandwiches for lunch on 70% of school days, otherwise I have a school meal.

If I have sandwiches, the probability that I buy a drink from the canteen is 0·2.

If I have a school meal the probability that I buy a drink is 0·9.

Find the probability that I buy a drink.

Here the probabilities are given to us.

The tree diagram for this situation is like this.

The probability of buying a drink

= 0·7 × 0·2 + 0·3 × 0·9

= 0·14 + 0·27

= 0·41

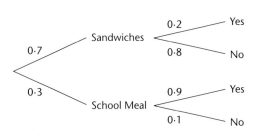

Dinner   Drink

EXAM TIPS

Some questions may instruct you to draw a tree diagram, others may not. Whilst it is not essential to draw a tree diagram for all questions it is a very powerful method for tackling probability problems. Even if you are not told to do so, you may find it helpful to draw one.

In problems involving selections from bags and so on, always assume that the probabilities are **dependent** unless you are specifically told that there is replacement or that the events are independent.

EXERCISE 9.2

1   There are 5 blue balls in a bag and 4 white ones.
If a white one is selected first, what is the probability that the second ball is blue?

2   500 tickets are sold in a raffle.
I buy 2 tickets.
I do not win the first prize.
What is the probability that I win the second prize?

3   Soraya and Robert are selecting cards in turn from an ordinary pack of 52 playing cards.
Soraya selects a heart and does not replace it.
What is the probability that Robert also selects a heart?

4   There are 80 tulip bulbs and 120 daffodil bulbs in a tub.
Alan picked 2 bulbs and there was 1 of each.
Charlie then picked a bulb.
What is the probability that Charlie's bulb was a tulip?

5 The probability that the school will win the first hockey match of the season is 0·6.

If they win the first match of the season the probability that they win the second is 0·7, otherwise it is 0·4.

a) Copy and complete the tree diagram.

1st match

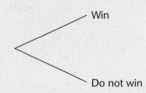

Win

Do not win

b) Find the probability that they win 1 of the first 2 matches only.

c) Find the probability that they win at least 1 of the first 2 matches.

6 There are 8 green balls in a bag and 2 white ones.

2 balls are selected at random without replacement.

a) Draw a tree diagram to show the probabilities of the possible outcomes.

b) Find the probability of selecting 2 white balls.

c) Find the probability of selecting 2 balls of different colours.

7 1000 tickets are sold in a raffle. I buy 2 tickets.

What is the probability that I win both of the first two prizes?

8 There are 7 yellow balls and 4 red balls in a bag.

2 balls are selected at random without replacement.

Find the probability of selecting at least 1 red ball.

9 There are 5 red balls, 3 yellow balls and 2 green balls in a bag. 2 balls are selected at random without replacement.

a) Find the probability of selecting 2 red balls.

b) Find the probability of selecting 2 balls of the same colour.

10 On average, Gary takes sandwiches to school for lunch on 2 days a week, otherwise he has a school meal.

If he takes sandwiches the probability that he has time to play football is 0·8.

If he has a school meal the probability that he has time to play football is 0·3.

Find the probability that on any given day Gary has time to play football.

11 Salima walks to school, cycles or goes by bus.

The probability that she walks is 0·5. The probability that she cycles is 0·2. If she walks, the probability that she is late is 0·2.

If she cycles, it is 0·1 and if she goes by bus, it is 0·4.

Find the probability that, on any given day, she is on time for school.

12 If Ryan is fit to play, the probability that United will win their next match is 0·9.

If he is not fit to play, the probability is 0·8.

The physiotherapist says there is a 60% chance he will be fit.

What is the probability that United will win their next match?

13 There are 3 red pens and 5 blue ones in a pencil case.

Jenny picked 2 pens at random.

Find the probability that Jenny picked at least 1 blue pen.

14 There are 5 grey socks, 3 black socks and 4 navy socks in Lisa's drawer. She selects 2 socks at random. What is the probability that she selects a pair of the same colour socks?

15 Sanjay selects 3 cards without replacement from a normal pack of 52 playing cards. What is the probability that he selects 3 aces?

16 There are 5 men and 4 women on a committee. 2 are selected at random to represent the committee on a working party. What is the probability that the two selected are
  a) both women?
  b) both men?
  c) one woman and one man?

17 Paul chooses 2 cards without replacement from an ordinary pack of 52 playing cards. Find the probability that he chooses
  a) 2 hearts.
  b) 2 kings.
  c) a king and a queen.

18 There are 7 red balls and 3 blue ones in a bag. Rosemary selects 3 balls at random without replacement.
  a) Draw a tree diagram to show the probabilities of the possible outcomes.
  b) Find the probability that Rosemary chooses at least 1 red ball.
  c) Find the probability that she chooses 2 blue balls and 1 red ball.

KEY IDEAS

- If events A and B are mutually exclusive, then P(A or B) = P(A) + P(B).

- P(A and B) = P(A) × P(B)

- If the outcome of the second event is affected by the outcome of the first event, the probability of the second event will vary according to what happens in the first event. These events are called dependent events.

- When events are dependent, the probability of the second event is called a conditional probability since it is conditional on the outcome of the first event.

- Tree diagrams can be used when working with conditional probabilities.

STAGE
9

Working in two and three dimensions

You will learn about

- Calculating the distance between two points
- Using right-angled triangles to find lengths and angles in three dimensions
- Finding angles between lines and planes

You should already know

- How to use coordinates in two and three dimensions
- How to apply Pythagoras' theorem
- The convention for labelling sides and angles in a triangle
- How to use trigonometry in right-angled triangles

The distance between two points

You can use Pythagoras' theorem to find the distance between two points on a graph.

EXAMPLE 1

Find the length AB.

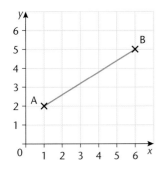

First make a right-angled triangle by drawing across from A and down from B.

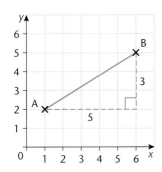

You can see that the lengths of the short sides are 5 and 3.

You can then use Pythagoras' theorem to work out the length of AB.

$AB^2 = 5^2 + 3^2$
$AB^2 = 25 + 9$
$AB^2 = 34$
$AB = \sqrt{34}$
$AB = 5 \cdot 83$ units to 2 decimal places

STAGE
9

EXAMPLE 2

A is the point ($^-$5, 4) and B is the point (3, 2).

Find the length AB.

Plot the points and complete the right-angled triangle.

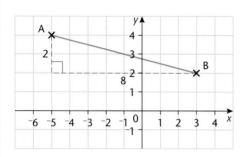

Then use Pythagoras' theorem to work out the length of AB.

$$AB^2 = 8^2 + 2^2$$

$$AB^2 = 64 + 4$$

$$AB^2 = 68$$

$$AB = \sqrt{68}$$

$$AB = 8 \cdot 25 \text{ units to 2 decimal places}$$

It is also possible to answer Example 2 without drawing a diagram.

From A($^-$5, 4) to B(3, 2) the *x*-value has been increased from $^-$5 to 3.
That is, it has been increased by 8.

From A($^-$5, 4) to B(3, 2) the *y*-value has been decreased from 4 to 2.
That is, it has been decreased by 2.

So the short sides of the triangles are 8 units and 2 units.

As before, you can now use Pythagoras' theorem to work out the length of AB.

You may prefer, however, to draw the diagram first.

EXERCISE 10.1

1 Find the length of each of the lines in the diagram.
Where the answer is not exact, give your answer correct to 2 decimal places.

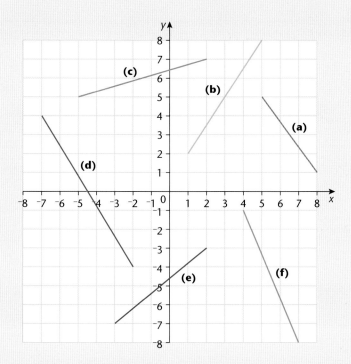

2 Find the length of the line joining each of these pairs of points.
You can draw a diagram to help you.
Where the answer is not exact, give your answer correct to 2 decimal places.
a) A(1, 4) and B(6, 8)
b) C(1, 7) and D(6, 2)
c) E(4, 1) and F(6, ⁻3)
d) G(⁻3, ⁻4) and H(0, 2)
e) I(⁻5, ⁻1) and J(1, 1)
f) K(⁻5, 0) and L(0, 12)

Finding lengths and angles in three dimensions

You can find lengths and angles of three-dimensional objects by identifying right-angled triangles within the object and using Pythagoras' theorem or trigonometry.

EXAMPLE 3

The rectangular box measures 4 cm by 3 cm by 6 cm.

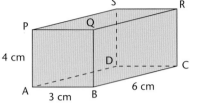

Calculate these.

a) AC   **b)** AR

c) Angle RBC   **d)** Angle ARC

a) $AC^2 = AB^2 + BC^2$

$AC^2 = 3^2 + 6^2$

$AC^2 = 45$

$AC = 6.71$ cm to 2 decimal places

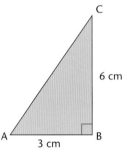

b) $AR^2 = AC^2 + RC^2$

$AR^2 = 45 + 4^2$

$AR^2 = 61$

$AR = 7.81$ cm to 2 decimal places

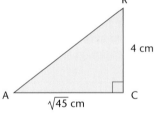

> **EXAM TIP**
>
> Don't use 6.71^2 – this has been rounded. You need AC^2 and this is 45.

c) $\tan x = \dfrac{4}{6}$

$x = \tan^{-1}\dfrac{4}{6}$

$x = 33.7°$ to 1 decimal place

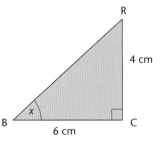

> **EXAM TIP**
>
> When there is a choice of trig. formulae to use, use the one which contains as many given values as possible. You may have calculated a value incorrectly.

d) $\tan x = \dfrac{\sqrt{45}}{4}$

$x = \tan^{-1}\left(\dfrac{\sqrt{45}}{4}\right)$

$x = 59.2°$ to 1 decimal place

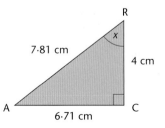

> **EXAM TIP**
>
> You can't avoid using a calculated value here. Using 6.71 will give an answer correct to 3 significant figures but it is good practice to use the more accurate value, $\sqrt{45}$, and round at the end.

In Example 3, notice how you used the result of part **a)** to work out the result of part **b)**. You can use a similar method to work out the length of the diagonal of a cuboid measuring a by b by c.

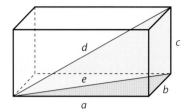

The diagonal of the rectangular base is labelled e.

$e^2 = a^2 + b^2$

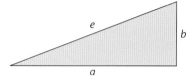

The diagonal of the cuboid is labelled d.

$d^2 = e^2 + c^2$

But $e^2 = a^2 + b^2$ so

$d^2 = a^2 + b^2 + c^2$

$d = \sqrt{a^2 + b^2 + c^2}$

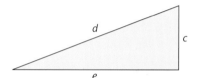

▌| EXAMPLE 4

Find the length of the diagonal of this cuboid.

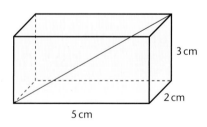

The length of the diagonal of a cuboid $= \sqrt{a^2 + b^2 + c^2}$.

In this cuboid $a = 5\,\text{cm}$, $b = 2\,\text{cm}$ and $c = 3\,\text{cm}$.

Length of diagonal $= \sqrt{5^2 + 2^2 + 3^2}$

$\qquad\qquad = \sqrt{25 + 4 + 9}$

$\qquad\qquad = \sqrt{38}$

$\qquad\qquad = 6\cdot2\,\text{cm}$ to 1 decimal place

STAGE
9

EXAMPLE 5

A tree, TC, is 20 m north of point A.

The angle of elevation of the top of the tree, T, from A, is 35°.

A point, B, is 30 m east of point A.

A, B and C are on horizontal ground.

Calculate
a) The height of the tree, TC.
b) The length BC.
c) The angle of elevation of T from B.

First, a diagram of the situation is needed.

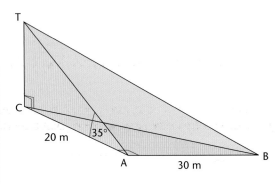

a) $\tan 35° = \dfrac{TC}{20}$

   $TC = 20 \tan 35°$

   $TC = 14 \cdot 0$ m to 1 decimal place

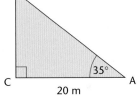

b) $BC^2 = 20^2 + 30^2$

   $BC^2 = 1300$

   $BC = 36 \cdot 1$ m to 1 decimal place

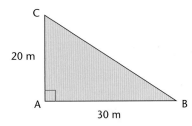

c) $\tan x = \dfrac{14 \cdot 0 \ldots}{36 \cdot 1 \ldots}$

   $x = \tan^{-1}\left(\dfrac{14 \cdot 0 \ldots}{36 \cdot 1 \ldots}\right)$

   $x = 21 \cdot 2°$ to 1 decimal place

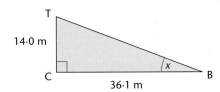

EXAMPLE 6

A mast, MG, is 50 m high.

It is supported by two ropes, AM and BM, as shown.

ABG is horizontal.

Other measurements are shown on the diagram.

Is the mast vertical?

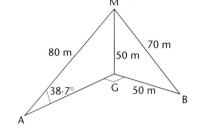

If angle MGA is a right angle then $\sin 38.7° = \dfrac{50}{80}$.

$\sin^{-1}\left(\dfrac{50}{80}\right) = 38.68 \ldots$

So angle MGA is a right angle.

Is angle MGB a right angle?

$50^2 + 50^2 = 5000 \qquad \sqrt{5000} = 70.7\ldots$

MB is shorter than this, so angle MGB is not a right angle.

The mast leans towards B (angle MGB < 90°).

EXERCISE 10.2

1 ABCDEFGH is a rectangular box with dimensions as shown.

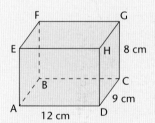

    a) Calculate these.
        (i) Angle GDC
        (ii) EG
        (iii) EC
        (iv) Angle GEC
    b) Take A as origin, AD as *x*-axis, AB as *y*-axis and AE as *z*-axis.
        Write down the coordinates of the vertices.

2 A wedge has a rectangular base, ABCD, on horizontal ground. The rectangular face, BCEF, is vertical.

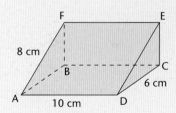

    Calculate these.
    a) FC
    b) Angle DFC
    c) FD
    d) Angle EDF

STAGE

9

Working in two and three dimensions

3 Three points, A, B and C, are on horizontal ground with B due west of C and A due south of C.
A chimney, CT, at C, is 50 m high.
The angle of elevation of the top, T, of the chimney is 26° from A and 38° from B.
Calculate these.
a) How far A and B are from C
b) The distance between A and B
c) The bearing of B from A

4 In the cuboid, PQ = 7·5 cm, QR = 4 cm and PX = 12 cm.

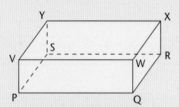

Calculate these.
a) XR
b) Angle PXR

5 The diagram shows a square-based pyramid with V vertically above the centre, X, of the square ABCD.
AB = 8 cm and AV = 14 cm.

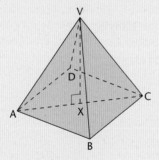

Calculate these.
a) Angle CAB
b) Angle VCB
c) AC
d) AX
e) VX
f) Angle VAX

Hint:
Triangle VCB is isosceles

6 PQRSTU is a triangular prism.
PQ = SR = TU = 5·7 cm.
PT = QU = 4·3 cm, PR = 6·9 cm.

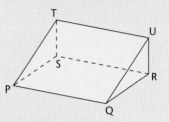

Calculate these.
a) PU
b) QR
c) UR = 2·1 cm.
Is angle URQ a right angle?
Show how you decide.

7 ABCDPQ is a triangular prism with ABPQ horizontal and ADP vertical.

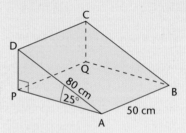

Calculate these.
a) DP
b) AC
c) Angle CAQ
d) The volume of the prism

8 A pyramid has a rectangular base, ABCD, with AB = 15 cm and BC = 8 cm.
The vertex, V, of the pyramid is directly above the centre, X, of ABCD with VX = 10 cm.
a) Calculate these.
(i) AC
(ii) AV
(iii) Angle AVB
b) M is the midpoint of AB.
Calculate these.
(i) VM
(ii) Angle VMX

STAGE
9

9 ABCDEFGH is a cuboid.
AB = 7 cm, AC = 8·6 cm and angle
GBC = 41°.

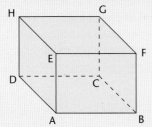

Calculate these.
a) BC
b) Angle GAC

10 Triangle ABC is horizontal.
X is vertically above A.
AC = AX = 15 cm.
Angle ACB = 27° and angle BAC = 90°.

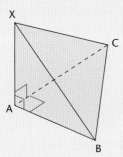

Calculate these.
a) XC
b) BC
c) BX

11 The diagram shows a garden shed with
the floor horizontal and the walls
vertical.

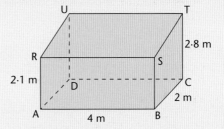

a) Calculate these.
(i) AT
(ii) ST
(iii) The angle that the roof of the
shed makes with the
horizontal
(iv) RT

b) Take A as the origin, AB as the
x-axis, AD as the y-axis and AR as
the z-axis.
Write down the coordinates of the
vertices.

12 A field, ABCD, is a quadrilateral with
opposite sides equal.
AD = BC = 80 m.
DC = AB = 35 m.
A vertical post, CE, is at one corner of
the field. The angles of elevation of
the top of the post are 7·8° from A
and 8·5° from B.

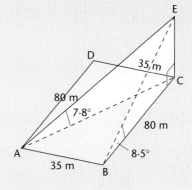

Is the field a rectangle?
Show how you decide.

STAGE
9

EXERCISE 10.2 continued

13 The pyramid OABCD has a horizontal rectangular base ABCD as shown.
O is vertically above A.

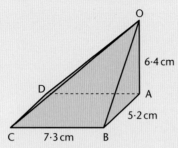

6·4 cm

5·2 cm

7·3 cm

Calculate these.
a) The length of OC
b) Angle OCA

14 A vertical mast, MT, has its foot, M, on horizontal ground.
It is supported by a wire, AT, which makes an angle of 65° with the horizontal and is of length 12 m, and two more wires, BT and CT, where A, B and C are on the ground.
BM = 4·2 m.

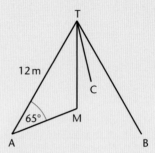

12 m

65°

Calculate these.
a) The height of the mast
b) The angle which BT makes with the ground
c) The length of wire BT

Three-dimensional coordinates

In Stage 7 you learned about three-dimensional coordinates.

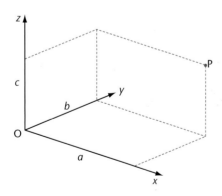

The coordinates of P are (a, b, c).

You can find the distance between two points in three dimensions by thinking of them as the opposite vertices of a cuboid and using the formula for the length of the diagonal of a cuboid, $\sqrt{a^2 + b^2 + c^2}$.

EXAMPLE 7

Calculate the distance between the points (⁻1, 0, 3) and (2, 4, 7).

To get from the first point to the second you need to move 3 units in the positive x-direction, 4 units in the positive y-direction and 4 units in positive z-direction.

Think of the two points as being at the vertices of a 3 by 4 by 4 cuboid.

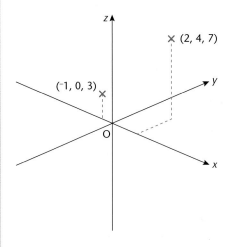 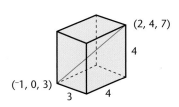

So the distance between the points (⁻1, 0, 3) and (2, 4, 7) is

$$\sqrt{3^2 + 4^2 + 4^2} = \sqrt{9 + 16 + 16}$$

$$= \sqrt{41}$$

$$= 6 \cdot 4 \text{ units to 1 decimal place}$$

EXERCISE 10.3

1   Calculate the distance between each of these pairs of points.
   a)   (4, 1, 5) and (2, 4, ⁻5)
   b)   (⁻2, 3, 6) and (2, 4, ⁻4)
   c)   (0, 2, ⁻3) and (1, 5, 12)
   d)   (0, 1, 4) and (2, 6, 5)
   e)   (3, 2, ⁻4) and (⁻1, 6, 7)

2   The distance of the point (2, 5, k) from the origin is $\sqrt{65}$.
   Find the value of k.

STAGE
9

The angle between a line and a plane

In the previous exercises of this chapter, you have been finding the angle between a line and a plane when the position of three points forming the angle have been specified. This next section deals with finding the angle when only the line and the plane are given.

The end, X, of the line XY is on the plane ABCD. The angle between the line and the plane is the angle YXP, where P is the point on the plane vertically below Y.

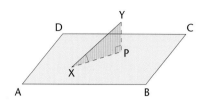

EXAMPLE 8

For this triangular prism, sketch the triangle and label the angle between the line and the plane given.

a) DY and ABCD **b)** AY and ABCD

c) AY and BCYX **d)** BY and ABX

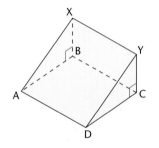

a) Angle YDC

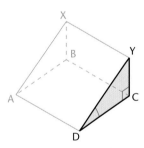

b) Angle YAC

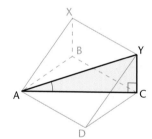

c) Angle AYB

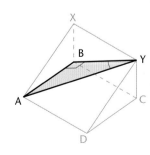

d) Angle YBX

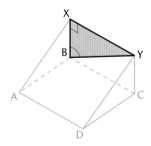

EXAMPLE 9

ABCDVWXY is a cuboid.

Calculate the angle between the following lines and planes.

a) DW and ABCD

b) DW and ABWV

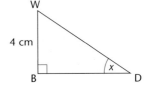

a) Angle BDW is the required angle.

First find length BD.

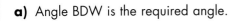

$BD^2 = 6^2 + 3^2$

$BD^2 = 45$

$BD = \sqrt{45}$ cm

$\tan x = \dfrac{4}{\sqrt{45}}$

$x = \tan^{-1}\left(\dfrac{4}{\sqrt{45}}\right)$

$x = 30.8°$ to 1 decimal place

b) Angle DWA is the required angle.

First find length WA.

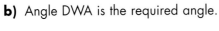

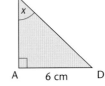

$WA^2 = 3^2 + 4^2$

$WA^2 = 25$

$WA = 5$ cm

$\tan x = \dfrac{6}{5}$

$x = \tan^{-1}\left(\dfrac{6}{5}\right)$

$x = 50.2°$ to 1 decimal place

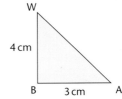

STAGE

9

1 The diagram shows the cuboid ABCDEFGH.

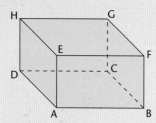

For this cuboid, sketch the triangle and label the angle between these lines and planes.
a) EB and ABCD
b) EB and ADHE
c) AG and ABCD
d) AG and CDHG

2 Given that, for the cuboid in question **1**, AB = 8 cm, BC = 6 cm and GC = 5 cm, calculate the angles between the lines and the planes listed above.

3 The diagram shows the tetrahedron ABCD.
ABC is a right-angled triangle on a horizontal plane.
D is vertically above A and angle DBC = 90°.

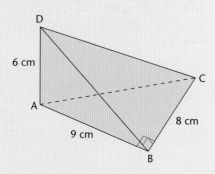

Calculate these.
a) The angle between BD and ABC
b) The angle between DC and ABD

4 In the pyramid, ABCD is a rectangle and V is vertically above the centre of the rectangle.
M is the midpoint of AD.

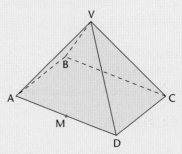

Sketch the triangle and label the angle between these lines and planes.
a) VC and ABCD
b) VM and ABCD

5 Given that, for the pyramid in question **4**, AD = 8 cm, DC = 6 cm and that VA = VB = VC = VD = 12 cm, calculate the angles between the lines and the planes listed above.

6 The diagram shows a tetrahedron.
AB = AC = BC = 4 m and VA = VB = VC = 6 m.

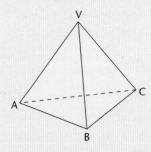

Calculate the angle that line VA makes with plane ABC.

Hint:
V lies directly above a point, X, which is $\frac{2}{3}$ of the way from A, along the bisector of angle CAB.

7 A cuboid has a base of sides 5·6 cm and 8·2 cm. Its height is 4·3 cm.
Calculate the angle between a diagonal of the cuboid and
  a) the base.
  b) a 5·6 cm by 4·3 cm face.

8 A pyramid is 8 cm high and has a square base of side 6 cm.
Its sloping edges are all of equal length.
Calculate the angle between a sloping edge and the base.

9 A square-based pyramid has height 8·3 cm.
Its sloping faces each make an angle of 55° with the base.
  a) Find the length of the sides of the square.
  b) Find the length of the sloping edges of the pyramid.

10 ABCDEF is a triangular wedge.
The base ABFE is a horizontal rectangle.
C is vertically above B.

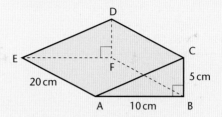

  a) Calculate the length AD.
  b) Calculate the angle which AD makes with
    (i) the base ABEF.
    (ii) the face ABC.

11 A cube has side 10 cm.
Calculate the angle between a diagonal of the cube and a face of the cube.
Explain why your answer is true for all cubes.

K KEY IDEAS

- The distance between two points can be found using Pythagoras' theorem.

- Pythagoras' theorem states that in triangle ABC $c^2 = a^2 + b^2$.

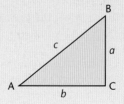

- The length of the diagonal of a cuboid with dimensions $a \times b \times c$ is $\sqrt{a^2 + b^2 + c^2}$.

- When working in three dimensions, identify the triangle containing the unknown side or angle and sketch it out separately. Then use Pythagoras' theorem and trigonometry as appropriate.

- $\sin x = \dfrac{\text{opposite}}{\text{hypotenuse}}$ $\cos x = \dfrac{\text{adjacent}}{\text{hypotenuse}}$ $\tan x = \dfrac{\text{opposite}}{\text{adjacent}}$

Working in two and three dimensions

STAGE
9

Revision exercise C1

1 a) Two triangles are similar. One has a base 6 cm long and an area of 20 cm². The other has a base 18 cm long. What is its area?

b) A cone has a height of 12 cm, a surface area of 380 cm² and a volume of 480 cm³. A similar cone has a volume of 60 cm³. What are its height and surface area?

c) Three similar cylindrical cans have diameters of 5 cm, 8 cm and 10 cm. The smallest can holds 150 ml. What is the capacity of each of the other two cans?

2 Two bottles are similar. Their heights are in the ratio 1 : 1·5. The larger one holds 2700 ml. What does the smaller one hold?

3 The areas of two similar pieces of paper are in the ratio 1 : 8. The larger piece of paper is 21·0 cm wide. What is the width of the smaller piece?

4 Draw a set of axes with the x-axis from 0 to 16 and the y-axis from 0 to 8. Plot the points A(4, 5), B(9, 5), C(9, 2) and D(4, 2) and join them to form a rectangle. Enlarge the rectangle by a scale factor of ⁻1·5 using the point (8, 4) as the centre of enlargement. Write down the coordinates of the enlarged rectangle.

5 Copy the diagram and find
a) the centre of enlargement.
b) the scale factor.

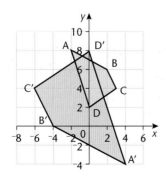

6 The probability that Zoe is late for school is 0·2. The probability that she is on time for school is 0·65. What is the probability that Zoe is either on time or late for school?

7 Assuming that the events in question **6** are independent, find the probability that Zoe is late for school on Monday and on time for school on Tuesday.

8 The probability that it is sunny on any day in January is 0·3. Find the probability that for two days in January
a) both are sunny.
b) one of the days is sunny.

9 The probability that the school netball team will win any match is 0·4. The probability that they draw any match is 0·1. The team play two matches. Find the probability that the team
a) loses both matches.
b) does not lose both matches.
c) wins one of the two matches and draws the other.

10 The probability that it rains on 15 July is 0·1. The probability that it rains on 16 July is also 0·1. Find the probability that it
a) rains on both days.
b) rains on one of the two days.

11 If it is fine when he gets up there is a probability of 0·7 that Richard will cycle to school. If it is raining when he gets up there is a probability of 0·05 that he cycles. The weatherman estimates that there is a 20% chance that it will rain tomorrow morning. Using the weatherman's estimate, find the probability that Richard will cycle to school tomorrow.

12 In a game I toss a coin and spin one of these fair spinners.

If I toss a head I spin the five-sided spinner.
If I toss a tail I spin the six-sided spinner.
I need a 5 to win the game.
What is the probability that I win the game?

13 Find the length of the line joining each of these pairs of points.
Give your answers correct to 2 decimal places.
  a) A(2, 2) and B(4, 7)
  b) C($^{-}$2, 9) and D(5, 3)

14 A street light suspended 7·5 m above the ground illuminates a circle with circumference 30·8 m.

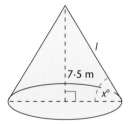

Calculate
  a) the angle marked $x°$.
  b) the length l.

15 In the diagram, triangle ABC is horizontal with angle ABC = 90°.
XB is vertical.

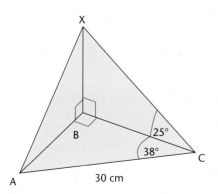

Find the length of XB.

16 Calculate the distance between each of these pairs of points.
  a) (0, 1, 4) and (2, 6, 5)
  b) (3, 2, $^{-}$4) and ($^{-}$1, 6, 7)
  c) ($^{-}$1, $^{-}$2, 1) and (3, 10, 4)

17 ABCDEFGH is a cuboid.

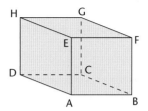

Specify the letters which define the angle between
  a) the line BH and the plane DCGH.
  b) the line AG and the plane BCGF.

18 A classroom measures 10 m by 7 m by 4 m.

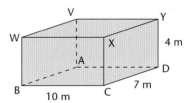

Calculate
  a) the angle between BY and ABWV.
  b) the angle between AC and ADYV.
  c) the length BY.

19 A glider, G, is 3 km east and 5 km south of its landing strip, L.
It is at an altitude of $\frac{1}{2}$ km above a point X on the ground.

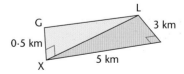

Calculate
  a) the bearing of the point X from L.
  b) the distance the glider has to fly to reach the landing strip.
  c) the glider's inclination as measured from L.

STAGE
9

11 Histograms

Drawing histograms

ACTIVITY 1

This frequency table shows the earnings in one week of a group of students.

Earnings (£w)	Frequency
$0 < w \leqslant 20$	5
$20 < w \leqslant 40$	15
$40 < w \leqslant 60$	26
$60 < w \leqslant 80$	30
$80 < w \leqslant 100$	6

Work in pairs, with one drawing a bar chart to represent these data and the other drawing a frequency polygon.

Compare your graphs and discuss the advantages and disadvantages of each type of graph.

A ACTIVITY 1 continued

The same data could have been grouped differently and presented like this.

Earnings (£w)	Frequency
$0 < w \leqslant 10$	2
$10 < w \leqslant 30$	8
$30 < w \leqslant 50$	24
$50 < w \leqslant 70$	32
$70 < w \leqslant 100$	16

How could you present this information on a graph?

Try out your ideas and discuss the results.

You may have realised in Activity 1 that our eyes take area into account when judging relative sizes. So, when the widths of the groups are unequal, you need to consider the impact of the area of the bars in a bar chart.

A histogram is a bar chart which uses the *area* of each bar to represent frequency. By doing this, it represents fairly the frequency of groups of unequal widths.

Histograms and bar charts are closely related.

- In a histogram the frequency of the data is shown by the area of each bar. (In a bar chart the frequency is shown by the height of each bar.)
- In a histogram the data are continuous. (Bar charts can be for discrete data.)
- Histograms have bars, or columns, whose width is in proportion to the size of the group of data each bar represents – the class width – so the bars may have different widths. (In a bar chart the widths of all the bars are usually the same.)
- In a histogram the vertical scale is the **frequency density**.
  Frequency density = frequency ÷ class width. (In a bar chart the vertical scale is the actual frequency.)

The table shows the ages of people in a netball club.

Age in years	Frequency
11–15	7
16–18	10
19–24	15
25–34	20
35–49	12
50–64	7

The histogram below shows the distribution of the ages of the members. Note that the upper boundary of the 11–15 age group is the 16th birthday. So the boundaries of the bars in the histogram are at 11, 16, 19, 25, 35, 50 and 65.

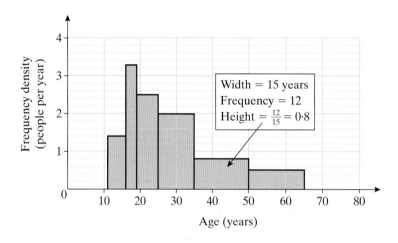

The calculation of the frequency density for one group has been added to the histogram to show you how it was done. Sometimes, instead of frequency densities being shown, a key is given, showing what each unit of area represents. On this histogram, one rectangle of the grid represents two people.

Usually the calculations for the frequency densities are done in a table, as in the next example.

EXAMPLE 1

An airline investigated the ages of passengers flying between London and Johannesburg.

The table shows the findings.

Draw a histogram to represent these data.

Age (A years)	Frequency
$0 \leq A < 20$	28
$20 \leq A < 30$	36
$30 \leq A < 40$	48
$40 \leq A < 50$	20
$50 \leq A < 70$	30
$70 \leq A < 100$	15

To draw a histogram you must first calculate the frequency density.

Age (A years)	Class width	Frequency	Frequency density
$0 \leq A < 20$	20	28	$28 \div 20 = 1\cdot4$
$20 \leq A < 30$	10	36	$36 \div 10 = 3\cdot6$
$30 \leq A < 40$	10	48	$48 \div 10 = 4\cdot8$
$40 \leq A < 50$	10	20	$20 \div 10 = 2$
$50 \leq A < 70$	20	30	$30 \div 20 = 1\cdot5$
$70 \leq A < 100$	30	15	$15 \div 30 = 0\cdot5$

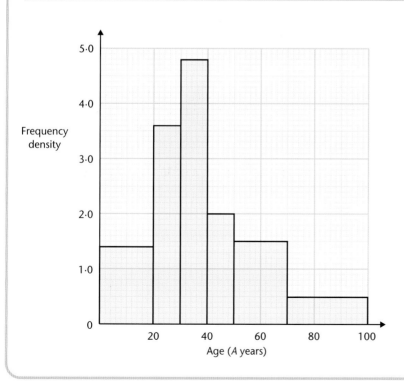

EXERCISE 11.1

1 200 commuters were surveyed to find the distances they travelled to work. The table shows the results.

Distance (*d* km)	Frequency
$0 \leq d < 5$	3
$5 \leq d < 10$	9
$10 \leq d < 15$	34
$15 \leq d < 20$	49
$20 \leq d < 30$	17
$30 \leq d < 40$	41
$40 \leq d < 60$	27
$60 \leq d < 100$	20

Draw a histogram to show this information.

2 The table shows the sizes of 48 marrows grown on an allotment.

Length (*L* cm)	Frequency
$0 \leq L < 20$	4
$20 \leq L < 40$	6
$40 \leq L < 50$	13
$50 \leq L < 60$	11
$60 \leq L < 90$	14

Show this information on a histogram.

3 The table shows the results of a survey to find the areas, to the nearest hectare, of 160 farms.

Area (*A* hectares)	Frequency
$1 \leq A < 4$	29
$4 \leq A < 8$	18
$8 \leq A < 12$	34
$12 \leq A < 16$	26
$16 \leq A < 24$	28
$24 \leq A < 30$	11
$30 \leq A < 34$	8
$34 \leq A < 40$	6

Draw a histogram to show this information.

4 The age of each person on a holiday coach tour is recorded. The table shows the results.

Age (*A* years)	Frequency
$0 \leq A < 10$	0
$10 \leq A < 20$	2
$20 \leq A < 30$	3
$30 \leq A < 45$	8
$45 \leq A < 50$	5
$50 \leq A < 70$	18
$70 \leq A < 100$	12

Draw a histogram to show this information.

5   A clothing manufacturer needs to know how long to make the sleeves of sweatshirts.
100 teenagers had their arm lengths measured.
The results are shown in the table.

Arm length (L cm)	Frequency
$40 \leq L < 45$	4
$45 \leq L < 50$	22
$50 \leq L < 55$	48
$55 \leq L < 60$	14
$60 \leq L < 70$	10
$70 \leq L < 80$	2

Draw a histogram to show this information.

6   An insurance company records the ages of people who were insured for holiday accidents etc. during a two-week period in August.

Age (A years)	Frequency
$0 \leq A < 5$	20
$5 \leq A < 10$	54
$10 \leq A < 20$	106
$20 \leq A < 30$	223
$30 \leq A < 40$	180
$40 \leq A < 60$	252
$60 \leq A < 90$	54

Draw a histogram to show this information.

7   The table shows the earnings for a group of students one week.

Earnings (£w)	Frequency
$0 < w \leq 20$	5
$20 < w \leq 40$	15
$40 < w \leq 70$	26
$70 < w \leq 100$	30
$100 < w \leq 150$	6

Draw a histogram to represent this distribution.
Label your vertical scale or key clearly.

8   This distribution shows the ages of people watching a local football team one week.

Age (years)	Frequency
Under 10	24
10–19	46
20–29	81
30–49	252
50–89	288

a) Explain why the class width of the 10–19 group is 10 years.
b) Calculate the frequency densities and draw a histogram to represent this distribution.

STAGE
9

Interpreting histograms

If the information is already given in a histogram it is possible to find out the frequencies and also to estimate the mean.

This is shown below.

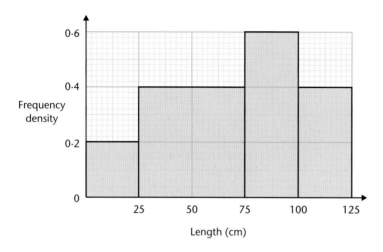

The frequency density = frequency ÷ width of group

so the frequency = frequency density × column width.

The mean can be estimated using the midpoint of each class.

Length	Frequency	Midpoint	Midpoint × frequency
$0 < x \leqslant 25$	5	12·5	62·5
$25 < x \leqslant 75$	20	50	1000
$75 < x \leqslant 100$	15	87·5	1312·5
$100 < x \leqslant 125$	10	112·5	1125
Total	50		3500

Estimate of mean $= \dfrac{3500}{50}$

$= 70$

EXERCISE 11.2

1 This histogram shows the ages of people who live in a small village.

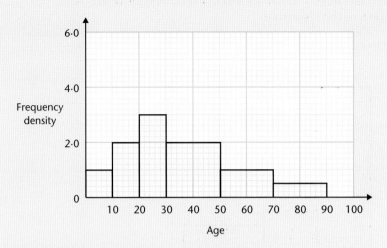

a) How many people live in the village?
b) Estimate their mean age.

2 This histogram shows the distribution of the weights of all the people living in a street.

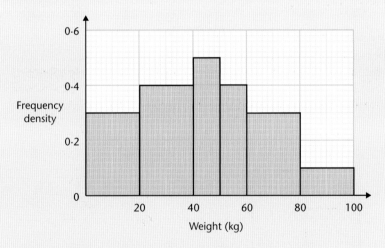

a) How many people live in the street?
b) Calculate an estimate of the mean weight.

3 A survey of students in a school was made to find the times it took them to travel from home to school each morning. The survey was made on one particular day.
The results are shown in the histogram.

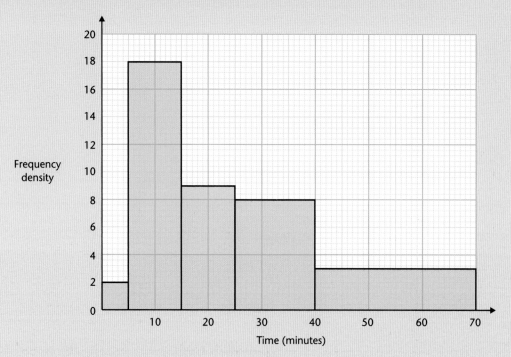

a) Find the number of students surveyed.
b) Calculate an estimate of the mean time it took to travel from home to school.

4 This histogram shows the results of a survey into the distance travelled by people to work each day.

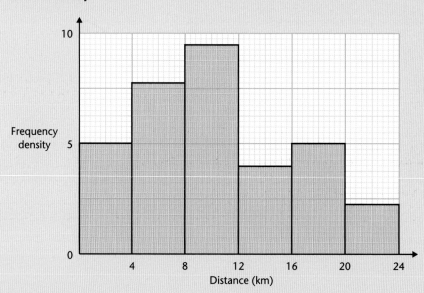

a) Find the number of people who were surveyed.
b) Calculate an estimate of the mean distance travelled to work.

5 The members of a gym were weighed.
The histogram represents their masses.

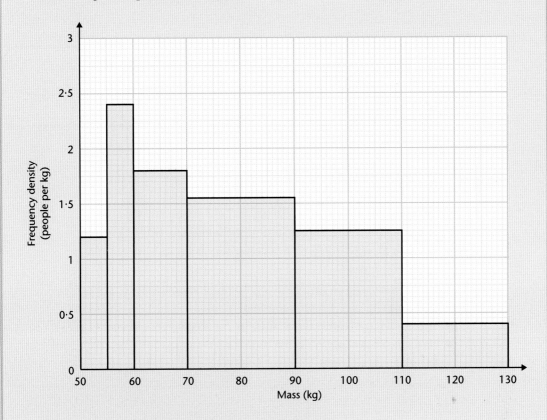

a) How many members of the gym were weighed?
b) Calculate an estimate of their mean mass.

STAGE
9

Circle properties

12

You will learn about

- Seven geometrical properties of a circle

You should already know

- That the sum of the angles in a triangle equals 180°
- That the exterior angle of a triangle equals the sum of the interior opposite angles
- That the sum of the angles on a straight line equals 180°
- The meaning of the term *congruent*
- The meaning of circle terms such as *arc*, *sector* and *segment*
- That the tangent at any point on a circle is perpendicular to the radius at the point

---

The angle subtended by an arc at the centre is twice the angle subtended at the circumference

BC is a chord of the circle whose centre is O.

Chord BC subtends (forms) angle BOC at the centre of the circle and angle BAC at the circumference.

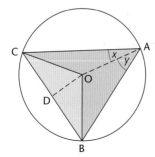

Here is a simple proof that angle COB = twice angle CAB.

Let angle CAO = x and angle BAO = y.

Then angle CAB = $x + y$.

Triangles OAB and OAC are isosceles   (OA, OB and OC are radii of the circle.)

Angle ACO = x and angle ABO = y (Base angles of isosceles triangles.)

Angle DOC = angle OAC + angle OCA = $2x$ (The exterior angle of a triangle equals the sum
Angle DOB = angle OAB + angle OBA = $2y$ of the opposite interior angles.)

Angle COB = angle DOC + angle DOB
            = $2x + 2y$
            = $2(x + y)$
            = $2 \times$ angle CAB

EXAM TIP

In a proof you should write out each step of your thinking and reasoning together with the reason or justification for each statement you write down.

What has just been proved can be expressed in words as

The angle at the centre of a circle = twice the angle at the circumference subtended by the same arc (or the same chord).

EXAMPLE 1

In this diagram O is the centre of the circle.

Calculate the value of angle a.

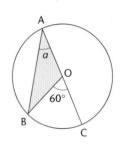

Angle $a = 30°$ (Angle at the centre = $2 \times$ angle at the circumference.)

STAGE
9

The angle subtended at the circumference in a semicircle is a right angle

The angle subtended at the circumference in a semicircle is a right angle. This is a special case of the fact that the angle subtended by an arc at the centre of a circle is twice the angle that it subtends at any point on the circumference.

The proof is given below but you should try to write out your own proof and then use this to check your work.

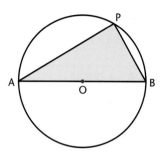

Angle AOB = 2 × angle APB   (The angle at the centre is twice the angle at the circumference.)

Angle AOB = 180°   (AB is a diameter, i.e. a straight line.)

Angle APB = 90°   (Half of angle AOB.)

So the angle in a semicircle is 90°.

This theorem is often used in conjunction with the angle sum of a triangle, as is shown in the next example.

▌▌ EXAMPLE 2

Work out the size of angle *x*.

Give a reason for each step of your work.

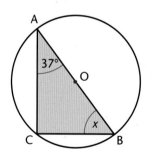

Angle ACB = 90°   (The angle in a semicircle is a right angle.)
    x = 180 − (90 + 37)   (The angle sum of a triangle is 180°.)
    x = 53°

STAGE
9

EXERCISE 12.1

In each of the following diagrams O is the centre of the circle.
Find the size of each of the lettered angles.
Write down each step with the reasons for your deductions.

1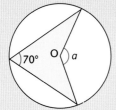

2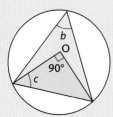

3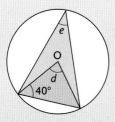

4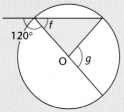

5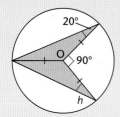

6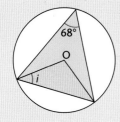

7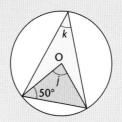

8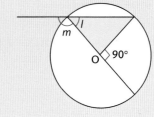

9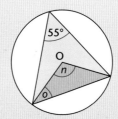

10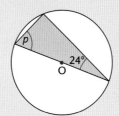

11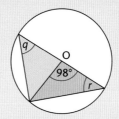

12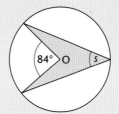

STAGE
9

13
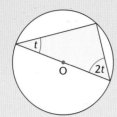

14
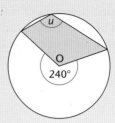

Angles in the same segment of a circle are equal

You can use the fact that the angle at the centre is twice the angle at the circumference to prove that angles in the same segment of a circle are equal.

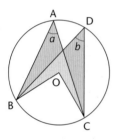

Angle $a = \frac{1}{2} \times$ angle BOC   (Angle at centre $= 2 \times$ angle at circumference.)

Angle $b = \frac{1}{2} \times$ angle BOC   (Angle at centre $= 2 \times$ angle at circumference.)

Therefore angle $a =$ angle b.

This well-known proof is often written as

The angles in the same segment of a circle are equal.

To use this property, you have to identify angles subtended by the same arc since these are in the same segment. This is shown in the next example.

EXAMPLE 3

Find the sizes of angles *a*, *b* and *c*.

Give a reason for each step of your work.

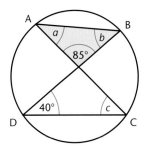

a = angle BDC = 40° (Angles in the same segment are equal.)

b = 180 − (40 + 85) = 55°   (The angle sum of a triangle is 180°.)

c = *b* = 55°   (Angles in the same segment are equal.)

The opposite angles of a cyclic quadrilateral add up to 180°

A cyclic quadrilateral is one with each of its vertices on the circumference of a circle. The proof that the opposite angles of a cyclic quadrilateral add up to 180° also relies on the proof that the angle at the centre is twice the angle at the circumference.

Again, you should try to write your own proof and then check it against the one given below.

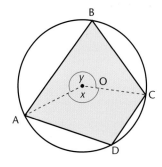

Angle ABC = $\frac{1}{2}x$ (The angle at the centre is twice the angle at the circumference.)

Angle ADC = $\frac{1}{2}y$ (The angle at the centre is twice the angle at the circumference.)

$x + y = 360°$ (The angles around a point add up to 360°.)

Angle ABC + angle ADC = $\frac{1}{2}(x + y)$

Angle ABC + angle ADC = $180°$

So opposite angles of a cyclic quadrilateral add up to 180°.

STAGE

9

The next example shows a typical application of the proof that the opposite angles of a cyclic quadrilateral add up to 180°.

EXAMPLE 4

Find the sizes of angles *c* and *d*.

Give a reason for each step of your work.

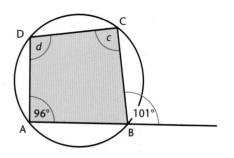

$c = 180 - 96 = 84°$ (Opposite angles of a cyclic quadrilateral add up to 180°.)

Angle ABC = $180 - 101 = 79°$ (Angles on a straight line add up to 180°.)

$d = 180 - 79 = 101°$ (Opposite angles of a cyclic quadrilateral add up to 180°.)

EXERCISE 12.2

In the following questions O is the centre of the circle.
Calculate the angles marked with letters.

1

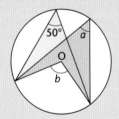

4

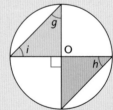

2

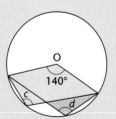

5

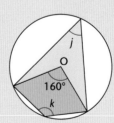

3

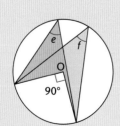

6

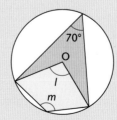

7

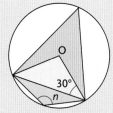

8

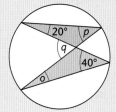

9

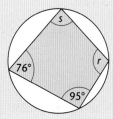

10

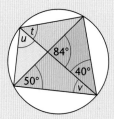

11

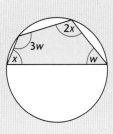

12

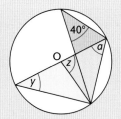

13

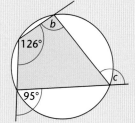

14

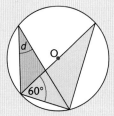

15

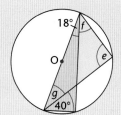

Sometimes the angle at the circumference is in the same semicircle as the arc.

EXAMPLE 5

Prove that angle QRS = $\frac{1}{2}$ × angle QOS.

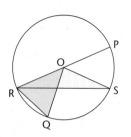

Let angle ORQ = x and angle ORS = y.

Triangle ORQ is isosceles.　　　　　　　(OR and OQ are radii.)

Therefore angle ROQ = 180° – 2x　　　(Angle sum of triangle.)

Triangle ORS is isosceles.　　　　　　　(OR and OS are radii.)

Therefore angle y = angle OSR

Therefore angle ROS = 180° – 2y　　　(Angle sum of triangle.)

Angle QOS = angle ROS – angle ROQ
　　　　　 = (180° – 2y) – (180° – 2x)
　　　　　 = 2x – 2y
　　　　　 = 2(x – y)

but angle QRS = angle ORQ – angle ORS
　　　　　　 = x – y

Therefore angle QOS = 2 × angle QRS.

You need to be able to identify when to use the circle theorems you have learned. Use this exercise to practise.

EXERCISE 12.3

Calculate the sizes of the angles marked with letters.
O is the centre of each circle.
Give the reasons for each step of your working.

1

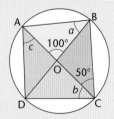

2

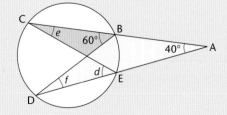

Circle properties

STAGE
9

12

120

3

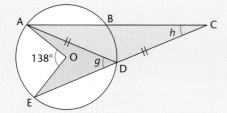

7

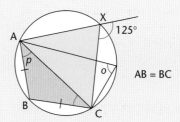

AB = BC

4

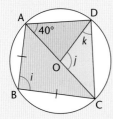

8

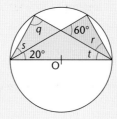

5

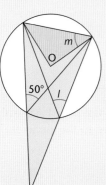

9

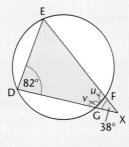

6

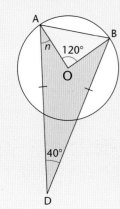

10

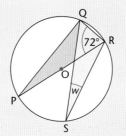

The perpendicular from the centre to a chord bisects the chord

AB is a chord to the circle centre O.

X is the midpoint of AB.

Triangle AOB is isosceles.

OX is a line of symmetry for this triangle.

Therefore angle OXA = angle OXB = 90°.

What has just been shown is

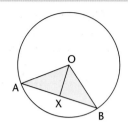

The straight line which joins the centre of a circle to the midpoint of a chord is at right angles to the chord.

An alternative way of stating this is

The perpendicular from the centre to a chord bisects the chord.

If the line OX is extended to become a radius, as in the diagram below, the chord AB will become the tangent at X. OX will be perpendicular to the tangent drawn to the circle at the point of contact, X.

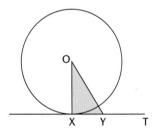

This fact is proved below.

Work through the proof and make sure you can follow it.

TX is the tangent to the circle centre O.

OX is the radius at the point of contact.

If angle OXT is not 90° it must be possible to draw a line at right angles to TX, the line OY, and so angle OYX = 90°.

If angle OYX = 90° then OX is the hypotenuse of triangle OYX and OX > OY.

What follows from this statement is that Y must be inside the circle because OX is a radius.

Therefore line TYX must cut the circle, i.e. line YX would be part of a chord.

This is impossible because TX is defined as a tangent.

Therefore it is impossible for angle OXT not to be 90°.

Therefore angle OXT is 90°.

This is an example of a **proof by contradiction**.

EXERCISE 12.4

In questions **1** to **8** calculate the angles marked with letters.
O is the centre of each circle.
X and Y are the points of contact of the tangents to each circle.

1

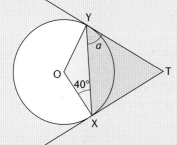

2

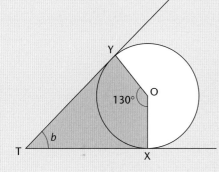

3

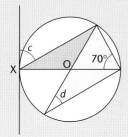

4

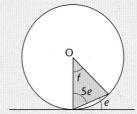

5

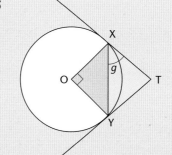

6

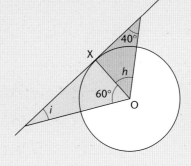

STAGE
9

EXERCISE 12.4 continued

7

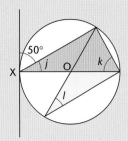

In questions **9** and **10**, find the lengths marked with letters.

9

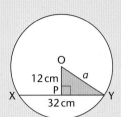

8

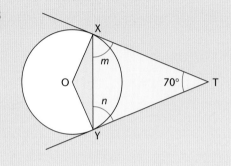

10

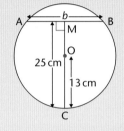

The two tangents to a circle from an external point are equal

T is a point outside a circle, centre O.

TX and TY are the tangents from T to the circle.

You can prove triangles OTX and OTY are congruent as follows.

$XT^2 = OT^2 - OX^2$ (Pythagoras)
$YT^2 = OT^2 - OY^2$ (Pythagoras)

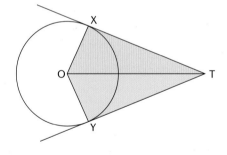

Therefore $XT^2 = YT^2$ and $XT = YT$.

Triangles OXT and OYT have corresponding sides equal and therefore corresponding angles equal.

OX = OY   (Radii)

Angles OXT and OYT are 90°   (Angle between radius and tangent.)

TO is common.

Therefore triangle OTX is congruent to triangle OTY.   (RHS)

If the triangles OTX and OTY are congruent then

XT = YT;   angle XTO = angle YTO;   angle TOX = angle TOY

You can state these facts in words.

> **Tangents drawn from a point to a circle are equal in length. They subtend equal angles at the centre of the circle and they make equal angles with the straight line joining the centre of the circle to the point.**

The angle between a tangent and a chord is equal to the angle in the alternate segment

BX is a chord of the circle centre O.

TX is a tangent meeting the circle at X.

AX is a diameter.

You can prove angle BXT = angle BAX as follows.

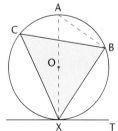

Angle ABX = 90°   (Angle in a semicircle.)

Angle OXT = 90°   (Angle between a diameter and a tangent.)

Therefore angle AXB = 90° – angle BXT

But angle AXB + angle BAX + angle ABX = 180°   (Angle sum of triangle.)

i.e. angle AXB + angle BAX = 90°

Therefore 90° – angle BXT + angle BAX = 90°

Therefore angle BXT = angle BAX

This is a general result since

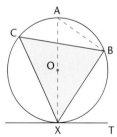

angle XCB = angle XAB (subtended by arc XB.)

You can state this fact in words.

> **The angle between a tangent and a chord drawn from the point of contact is equal to the angle subtended by the chord in the alternate segment, that is, on the other side of the chord.**

EXERCISE 12.5

In questions **1** to **12**, find the angles marked with letters.
Gives reasons for your answers.

1
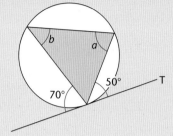

2
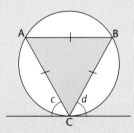

3
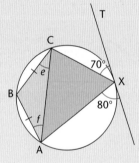

4
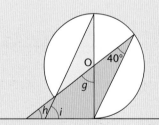

5
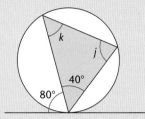

6
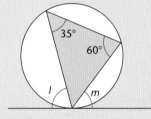

7
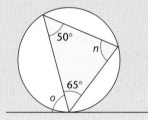

8
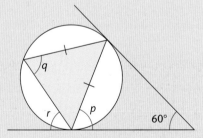

EXERCISE 12.5 continued

9

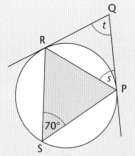

10

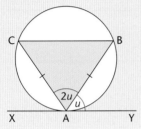

11

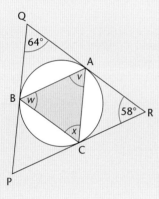

12

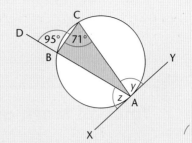

13 The sides of the quadrilateral TUVW are all tangents to the circle centre O. Prove that TU + WV = UV + WT.

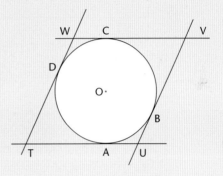

K **KEY IDEAS**

■ The angle subtended by an arc at the centre of a circle is twice the angle subtended at the circumference.

■ The angle subtended at the circumference in a semicircle is a right angle.

■ Angles in the same segment of a circle are equal.

■ The opposite angles of a cyclic quadrilateral add up to 180°.

■ The perpendicular from the centre to a chord bisects the chord.

■ The two tangents to a circle from an external point are equal.

■ The angle between a tangent and a chord is equal to the angle in the alternate segment.

13 Straight-line graphs

Finding the equation of a straight-line graph

In Stage 8 you learned that $y = mx + c$ is the equation of a straight line where m is the gradient of the line and c is the y-intercept.

You can use these facts to work out the equation of a line from its graph or, if you are given the equation, to find the gradient and where it crosses the y-axis without drawing the graph.

EXAMPLE 1

Find the equation of this straight line.

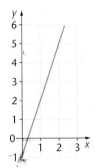

y = 3x − 1.

The gradient of a line, $m = \dfrac{\text{increase in } y}{\text{increase in } x}$

$$= \frac{6}{2}$$

$$= 3$$

The line passes through $(0, {}^{-}1)$ so the y-intercept, c, is $^{-}1$.

So the equation of the line is $y = 3x - 1$.

EXAMPLE 2

The equation of a line is $3x - 2y = 14$. *− 3x*

Find the gradient and the point where the line crosses the y-axis.

The equation must be in the form $y = mx + c$ so you need to rearrange the equation.

$$3x - 2y = 14$$
$$2y = 3x - 14$$
$$y = 1 \cdot 5x - 7$$

So the gradient is $1 \cdot 5$ and the point where the line crosses the y-axis is $(0, {}^{-}7)$.

STAGE
9

EXERCISE 13.1

1 Write down the equation of the straight line
   a) with gradient 3 and passing through (0, 2).
   b) with gradient ⁻1 and passing through (0, 4).
   c) with gradient 5 and passing through (0, 0).
   d) with gradient 4 and passing through (0, ⁻1).
   e) with gradient ⁻2 and passing through (0, 5).
   f) with gradient 3 and passing through the origin.

2 Find the equation of each of these lines.

a)

b)
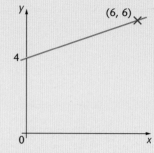

c)
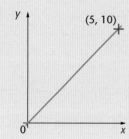

d)

e)

f)
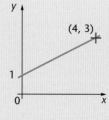

EXERCISE 13.1 continued

3 Find the equation of each of these lines.

a)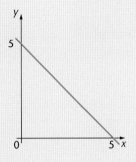

b)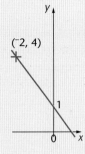

c)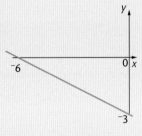

d)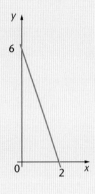

e)

f)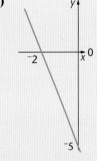

4 Find the gradient and y-intercept of each of these lines.
  a) $y = 4x - 6$
  b) $y = 7 - 9x$
  c) $y = x$
  d) $3y = x - 4$
  e) $x + 5y = 9$
  f) $4x + 2y = 11$

5 The table gives the cost when x metres of ribbon are sold.

Find an equation connecting x and C.

Number of metres (x)	0·25	0·5	1·75	3·00
Cost (C pence)	21	42	147	252

EXERCISE 13.1 continued

6 A shop which converts money offers these rates of dollars in exchange for pounds.

Pounds (p)	10	20	30
Dollars (d)	11	27	43

a) Draw a graph of these data.
b) Find the gradient of the graph and state what it represents.
c) What does the intercept on the d-axis represent?
d) Find the equation connecting d and p.

Parallel and perpendicular lines

ACTIVITY 1

a) Using graph-drawing software if possible, draw the following lines on separate pairs of axes.

Use values of x and y from $^-6$ to 6.

(i) $y = x$ and $y = {}^-x$

(ii) $y = 2x$ and $y = {}^-\frac{1}{2}x$

(iii) $y = 5x$ and $y = {}^-\frac{1}{5}x$

(iv) $y = 4x$ and $y = {}^-0.25x$

(v) $y = 2x$ and $x + 2y = 6$

b) What do you notice about each pair you have drawn?

Can you spot a connection between them?

Find more pairs that give the same result.

In Stage 8 you learned that parallel lines have the same gradient.

There is also a connection between the gradients of perpendicular lines.

STAGE
9

> If a line has gradient m, then a line perpendicular to it has gradient $\frac{^-1}{m}$.

An alternative way of stating this fact is

> The gradients of two perpendicular lines have a product of $^-1$.

EXAMPLE 3

The equation of a line is $y = 2x - 3$.

State the equation of a line

a) parallel to it.

b) perpendicular to it.

a) Any parallel line will have a gradient of 2.
One possible line is $y = 2x + 5$.

b) Any perpendicular line will have a gradient of $\frac{-1}{2}$.
One possible line is $y = \frac{-1}{2}x + 5$.

Notice that, as for parallel lines, the value of c can vary. It is the gradient that determines whether or not the lines are perpendicular.

You could rewrite this equation to get rid of the fraction.

$y = \frac{-1}{2}x + 5$ is the same as $2y = -x + 10$ or $2y + x = 10$ or $2y = 10 - x$.

EXERCISE 13.2

1 Draw the line $y = 2x$.
On the same graph, draw a line parallel to it and a line perpendicular to it.
Find the equations of the lines you have drawn.

2 Draw the line $y = -2x$.
On the same graph, draw a line parallel to it and a line perpendicular to it.
Find the equations of the lines you have drawn.

3 Draw the line $x + 3y = 6$.
State its gradient.
Draw a line perpendicular to this and state its gradient.

4 Draw the line $5x - y = 6$.
State its gradient.
Draw a line perpendicular to this and state its gradient.

5 Find the equation of the line parallel to $y = 6x - 3$ which passes through the point $(0, 4)$.

6 Find the equation of the line parallel to $y = 3x - 1$ which passes through the point $(0, 2)$.

7 Find the equation of the line perpendicular to $y = 0 \cdot 5x - 1$ which passes through the point $(0, 3)$.

8 Find the equation of the line perpendicular to $y = 0 \cdot 25x + 5$ which passes through the point $(0, -1)$.

9 A line joins $(1, 2)$ and $(5, -6)$.
Find the gradient of a line perpendicular to this.

STAGE

9

Straight-line graphs

10 A line joins ($^-$1, 2) and (1, 14). Find the gradient of a line perpendicular to this.

11 Find the equation of the line that passes through (1, 5) and is perpendicular to $y = 3x - 1$.

12 Find the equation of the line that passes through (0, 3) and is perpendicular to $2y + 3x = 7$.

13 Which of these lines are
   a) parallel?
   b) perpendicular?

   $y = 4x + 3$
   $2y - 3x = 5$
   $6y + 4x = 1$
   $4x - y = 5$

14 Two lines cross at right angles at the point (5, 3).
One of them passes through (6, 0).
What is the equation of the other line?

15 In the diagram AB and BC are two sides of a square, ABCD.

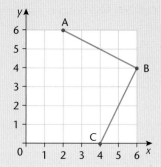

Work out
   a) the equation of the line AD.
   b) the equation of the line DC.
   c) the coordinates of D.

K **KEY IDEAS**

■ The equation of a line can be written as $y = mx + c$, where m and c are numbers.
m is the gradient of the line and c is the value of y where the graph crosses the y-axis.
In other words, the graph passes through (0, c).

■ Lines with the same gradient are parallel.

■ If a line has gradient m then a line perpendicular to it has gradient $\dfrac{^-1}{m}$.

STAGE
9

Surveys and sampling 14

Surveys

The diagram shows the four aspects of the data handling cycle.

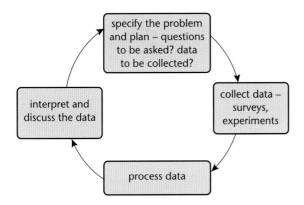

It is important to make sure that, in any investigation you carry out, the conclusions you make relate back to the initial aims. In other words, that you have answered the question and that you try to present a well-structured and coherent report.

STAGE
9

A ACTIVITY 1

a) Think back to a statistics investigation you have done in the past.
- What went well?
- What was difficult?
- Was there anything you wished afterwards that you had done differently?
- How can you apply what you learned then to your next project?

b) See how this fits into the data handling cycle.

A survey can be thought of as an investigation to establish something, for example,

- people's preferences, likes/dislikes or beliefs
- the number of plants or animals in a particular location
- the presence of oil or gas (a commercial example).

Remember that in an investigation you should explain how you are going to collect data, consider any practical problems in carrying out the survey, and explain how you will plan for these.

You should also use, and explain why you are using, a pilot survey or a pre-test of a questionnaire.

Questionnaire design

One common method of obtaining information from people is to use a questionnaire. Questionnaires need to be designed to provide answers that are easily analysed.

- Those with 'yes' or 'no' answers are clearly easy to analyse but remember that you must try to avoid writing questions which have an 'or' statement in them. For example, if you are doing a survey about the number of people who are vegetarian, asking a question such as 'Do you eat meat or vegetables?' will not provide any useful answers!

- One way is to ask questions giving options (boxes) from which the person selects a choice. This allows the information to be counted and analysed. Sometimes a box can be given as 'anything else' and a line left for comments, but it is a good idea to avoid this.

- Avoid questions which
  a) are not to the point (i.e. not about the subject)
  b) are embarrassing or biased.

If there was a general election tomorrow, how would you vote?

| Lab | Con | Lib | Green | Other |

How well do you consider the Prime Minister is running the country

| V. well | Well | No opinion | Badly | V. badly |

EXERCISE 14.1

1 Julie is designing a questionnaire for a survey about her local supermarket. She has decided to test two hypotheses.
   a) Local people visit the supermarket more often than people from further away.
   b) Local people spend less money per visit.
   Write two questions which could help her test each hypothesis.
   Each question should include at least three options for a response. The people interviewed are being asked to choose one of these options.

2 A Sports and Leisure Club offers a range of facilities for many different sporting activities. As part of their reviewing systems the committee issues questionnaires to a sample of members.

   One question said: 'If you are a regular tennis player and think better facilities are needed, how many extra courts should be provided and what other improvements would you like to see?'
   a) Write down at least two faults in this question.
   b) How would you improve the question? Write down your improved question.

Sampling

This section provides an introduction to sampling and the methods used to obtain a sample.

In data handling the word '**population**' is used for a collection, set or group of objects being studied.

A '**sample**' is a smaller group (a subset), selected from the population. If the population is large it is not usually possible or practicable to collect data on every member of that population, so one or more samples will be surveyed and conclusions will be drawn from these which will then be applied to the whole population.

In your work you should show that you are using a sample of an adequate size.

If the structure or composition of the population is known then it is important to ensure that the sample (or samples) represents that population, and thus that any variations in that population are reflected in the sample – which is therefore called a **representative sample**.

You must explain why you have chosen a particular sampling method, why the sample is of the size that it is, and also what effects the nature of the sample may have on your findings.

STAGE
9

There are various methods of choosing a representative sample.

1 Systematic random sampling

An example of this method would be the selection of a 10% sample by going through the population picking every tenth item or individual. The drawback is that this would only provide a representative sample if the population was arranged in a random way and not in a way that might introduce bias. (Bias is described later.)

2 Attribute random sampling

In this method the selection of the sample is made by choosing some attribute which is totally unrelated to the variable being investigated. Choosing a sample to investigate any relationship between head size and height from a list of people on the basis of their birthday being the first of the month would be an example of this.

3 Stratified random sampling

The population is divided into strata or subgroups and the sample is chosen to reflect the properties of these subgroups. For example, if the population contained three times as many people under 25 as over 25 then the sample should also contain three times as many people under 25. The sample should also be large enough for the results to be significant. This is sometimes called quota sampling.

4 Simple random sampling

However, if there is no information about the characteristics of the population – for example, no knowledge about the ages and sex of the people in the population, or about the colours and sizes of the objects in a population – then a sample must be selected on the basis that all items are equally likely to be chosen. This is called random sampling. To ensure a sample is random and as accurate as possible, ideally the sampling should always be repeated a number of times and the results averaged.

You will find examples of sampling taking place throughout the year. For example,

- in politics, with opinion polls reporting on the popularity of the political parties, especially in the weeks before local and general elections
- in market research – 'eight out of ten owners said their cats preferred Paws', or whether the building of the Millennium Dome was a good use of money, and so on.

You may well have seen market researchers interviewing people in the street.

A ACTIVITY 2

- Treat your class as a 'population' and calculate the average height exactly.

- Use each of methods 1 to 4 to identify a sample of about 6 to 8 students and see how close their average height is to the class average.

EXERCISE 14.2

1 A soap powder manufacturer wants to know what percentage of the population uses its washing powder. Would it be likely to obtain a representative sample by asking

a) people leaving a public house at 10 p.m. on a Friday night?

b) people leaving a supermarket at 7·30 p.m.?

c) people leaving a supermarket between 9 a.m. and noon each day for a week?

d) people getting off a commuter train on their way home from work?

Give reasons for each of your answers. Can you suggest a way of obtaining a representative sample?

2 You might be able to try the following experiment.

a) Put 100 counters in a bag. The counters should be the same size but of different colours (for example, some red, some blue).

b) Choose a colour (say, red).

c) Select ten counters without looking at them and make a note of the number of counters selected which are of your chosen colour (for example, three red ones). Return the counters to the bag.

d) Multiply by 10 to predict the number of red counters in the bag $(10 \times 3 = 30)$. (Write this number down so you don't forget it.)

e) Repeat steps **c)** and **d)** another nine times, giving a total of ten experiments (for example, you might get 20, 30, 40, 40, 50, 20, 30, 40, 40).

f) Find the mean of the ten experiments (for example, $\frac{340}{10} = 34$).

This figure ought to be close to the actual number of red counters in the bag. How does the answer you obtain compare with the actual number of counters of your chosen colour in the bag?

3 All schools could be improved. Write a questionnaire to give to a sample of students in your year or the whole school. Try to write questions which can be analysed easily. How are you going to select the sample?

4 Comment on the following ideas for obtaining a random sample. Give a better method if you can.

a) A random sample of all the adults in a town is required.
Method: Stand outside a supermarket and stop every tenth person who leaves between the hours of 9 a.m. and 3 p.m.

b) A random sample of the students at your school is required.
Method: Ask all the students to 'sign up' if they want to take part in a study and promise to pay £1 to all who are chosen. Choose at random until the required number of students is obtained.

5 Which of the following would you study by sampling?

a) The average life of a torch battery.

b) The top ten albums for, say, last month.

6 Which sampling method would you use in the following situations? Describe how to select a suitable sample in each case.

a) What proportion of the constituency you live in will vote for a particular party?

b) What is the average height of a student in your year group or school? Does the height of girls differ much from the height of boys?

c) What is the number of trees in a local wood?

d) What is the number of students in your school who would go to a school disco? Or a fair?

e) How much homework do students do? Does it increase as you get older?

Stratified random sampling

As mentioned earlier, sampling which is representative of the whole population is called stratified sampling. The method used is as follows.

- Separate the population into appropriate categories or strata, for example by age.
- Find out what proportion of the population is in each stratum.
- Select an appropriate number of members from each stratum of the population.
  This can be done by random sampling and so the technique is known as stratified random sampling.

EXAMPLE 1

The 240 students in Year 9 of a school are split into four groups for games. 90 play cricket, 70 play tennis, 30 choose athletics and the remaining 50 opt for volleyball.

Use a stratified random sample of 40 students to estimate the mean weight of all 240 students.

The sample size from each of the four groups must be in proportion to the stratum size, so the 40 students are selected as follows.

Cricket $\quad \dfrac{90}{240} \times 40 = 15$

Tennis $\quad \dfrac{70}{240} \times 40 = 11 \cdot 67$ i.e. 12

Athletics $\dfrac{30}{240} \times 40 = 5$

Volleyball $\dfrac{50}{240} \times 40 = 8 \cdot 33$ i.e. 8

Within each sample the actual students will be selected randomly (random sampling is discussed later).

The mean weights for each sample are found to be

Cricket	54·6 kg	Tennis	49·7 kg
Athletics	53·1 kg	Volleyball	47·9 kg

so the mean weight for all 240 students

$$= \frac{54 \cdot 6 \times 15 + 49 \cdot 7 \times 12 + 53 \cdot 1 \times 5 + 47 \cdot 9 \times 8}{40}$$

$= 51 \cdot 6$ kg to 1 decimal place

This is an estimate for the mean weight of the population of 240 students.

There is another form of sampling.

The natterjack toad is an increasingly threatened species of toad.

Scientists want to find out how many of these toads live in and around a pond.

To do this they catch 20 and mark them in a harmless way. The toads are then released.

Next day another 20 are caught: 5 of these toads have already been marked – in other words, a sample of 25% (5 out of a sample of 20) are marked. But 20 toads were marked initially.

This suggests that the original 20 toads were about 25% of the whole population.

$\frac{25}{100} \times P = 20$ therefore $P = 80$

Therefore the total population is 80.

EXERCISE 14.3

1   Amy wants to investigate the spending habits of students at her school. The number of students in each year group is as follows.

Year group	7	8	9	10	11
Number of students	208	193	197	190	184

Explain how Amy can obtain a stratified sample of 100 students for her survey.

2   The table shows the number of boys and girls in Year 10 and Year 11 of a school.

	Year 10	Year 11
Boys	120	134
Girls	110	100

The headteacher wants to find out their views about changes to the school uniform and takes a stratified random sample of 50 students from Year 10 and Year 11. Calculate the number of students to be sampled from Year 11.

STAGE
9

EXERCISE 14.3 continued

3 Scientists need to estimate the number of fish in Lake Hodder. They catch and 'tag' 450 fish and then release them back into the lake.
Over the next few days and at various locations they catch samples and count the number of fish that are tagged.

Day	Sample size	Number tagged
1	36	6
2	38	6
3	40	8
4	32	6

Use these values to estimate the total number of fish in the lake.

4 Repeat Exercise 14.2 question **2** (knowing the number of, say, red counters). Take four samples and use the number of red counters in each sample to estimate the total number of counters.
How close is this estimate to the known answer of 100?

Simple random sampling

Various methods are available in order to select the items for a random sample. These include

■ the random number facility included on scientific and graphical calculators – see your own calculator manual to find out how to use it to generate random numbers
■ using random number tables.

This is part of a random number table.

306377524300080380602156564232387322656656251899879221322606595460084874
04306678056086556696665100454587964544488455544850403456484565489785544
164512323588856898578846533165966623657526395268923325461548121532155465
553331548818416389615686645695696655484169631661564356153686631566463457
659894348793156936316633666316666459996963999733936315964065568625896255
489632258794521325952374123653586255625613545987533146332214896256589666
548965244888544752211323355232325456232566548795775229632214863221563321
5232325556232215625548563258984436544326542136545236

The following activity illustrates the method.

You might like to try this if you have time, either on your own or with friends. However, do work through the activity so that you understand the method even if you haven't time to do all the calculations.

A ACTIVITY 3

Plant Laboratories have produced a new variety of a flowering plant and are interested in the size of the flower heads. The diameters of the flowers on 50 plants are measured and shown as circles on the diagram below. The circles are numbered for reference.

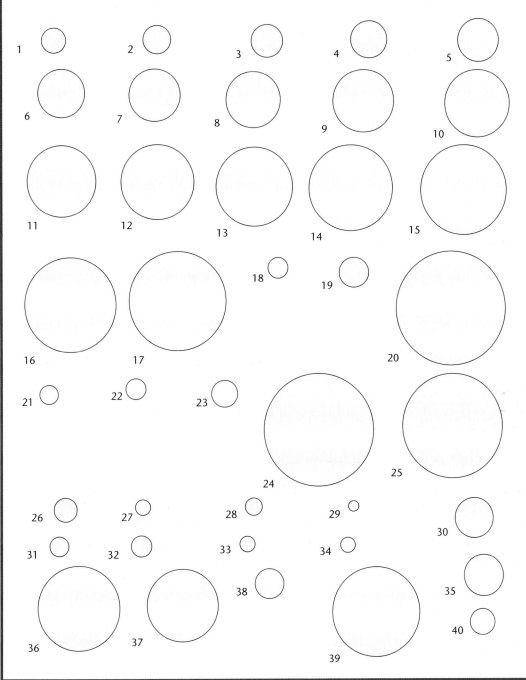

A ACTIVITY 3 continued

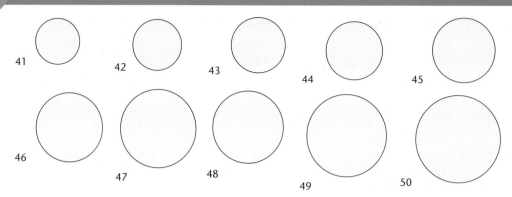

41 42 43 44 45

46 47 48 49 50

a) Select a sample of five circles that you think are representative. Measure their diameters and calculate the mean.

b) Repeat for another two samples and find the overall mean. (Collect the results from any others in your class who are also doing this activity.)

c) Using a random number table, such as the one on page 142,

    (i) choose a starting number in the table. Once you have fixed this you must move along the row or down the column and not 'jump about'. Divide the numbers into pairs, starting with your chosen number, to represent the two-digit reference numbers of the flowers, so 03 would represent picture 3. Discard or ignore any two-digit numbers that are greater than 50.

    (ii) write down the first five two-digit numbers that you get, measure the diameters of those circles and calculate the mean diameter.

    or use the random number generator on your calculator and follow steps **(i)** and **(ii)**.

d) The actual mean diameter for the 50 flowers is 1·43 cm.

    (i) How close were the means calculated by both methods?

    (ii) How did you judge 'representative'? In other words, what criteria did you use?

Discuss this with others in your class who have completed the experiment.

EXERCISE 14.4

1 The table shows the times taken by 100 students to solve a mathematics problem. The times are all in seconds.

14	20	73	35	28	39	25	17	16	23
20	7	13	30	39	36	17	35	57	26
150	39	25	27	25	40	39	62	47	25
22	16	32	46	29	21	57	10	122	81
90	34	12	68	28	81	32	47	35	37
39	40	23	46	25	43	74	53	24	51
12	30	93	26	17	21	32	37	33	42
93	40	18	55	11	56	34	67	13	15
104	21	25	49	35	18	15	47	26	57
38	92	59	12	32	46	36	25	71	35

Using a sample, estimate the mean and median times taken to solve the problem.
a) What sample size would you use and why?
b) Use your sample to obtain the estimates of the mean and median times.

2 Identify which type of sampling has been used in each of the following cases.
   a) In order to determine whether the sports facilities in a leisure centre are satisfactory, all the members are allocated a number and questionnaires are sent to 100 members, who are selected using random numbers.
   b) A car manufacturer employs 2000 people in the assembly plant, 400 in the parts department and 500 in the offices.
   A sample is taken consisting of 20 assembly plant workers, 4 people from the parts department and 5 from the offices.

3 A biologist wants to estimate the average number of worms per square metre in a field. The field is divided into 100 metre squares and each square has a two-digit identification number between 00 and 99.
   For example, square 56 (row 5, column 6) has 9 worms.

	0	1	2	3	4	5	6	7	8	9
0	5	6	4	7	4	7	6	4	8	3
1	2	6	4	8	4	6	5	6	6	3
2	7	8	4	3	2	7	8	9	3	3
3	6	4	5	5	8	3	2	1	4	6
4	8	6	3	6	4	5	6	3	7	8
5	4	3	4	6	5	8	9	2	4	3
6	7	6	4	1	2	3	5	5	4	3
7	3	7	6	4	2	3	6	3	4	2
8	6	5	5	6	3	4	3	3	6	7
9	5	4	7	6	9	4	3	5	6	4

You are going to select a sample of ten squares and find the mean number of worms in those squares, using the two methods below in turn.

Method 1: Starting at the beginning of the first row of the random number table on page 142, choose a simple random sample of ten squares.
Find the mean number of worms in these squares.

Method 2: Starting at the beginning of the second row of the random number table on page 142, choose the first value between 00 and 09.
Use this as the starting number of a systematic random sample of size 10.
Find the mean number of worms in these squares.

   a) Select the samples as described above and find the mean of each sample.
   b) How do the two samples compare?
   c) Find the mean of the population by finding the mean of all 100 squares.
   d) How do the means of the samples compare with the mean of the population of worms?

4 The table shows the number of students in Years 9 to 13 of a school.

Year 9	Year 10	Year 11	Year 12	Year 13
162	161	157	63	58

Students are to be interviewed about exam stress.
A stratified sample of 60 students is to be taken.
How many students from each of the year groups should be interviewed?

5 A company owns five factories. It wishes to select a stratified random sample of 100 of its workers to interview about a new pension scheme.

Factory	1	2	3	4	5
Number of workers	409	207	1985	1011	398

How many workers should be selected from each factory?

A ACTIVITY 4

Here is a list of investigations that could be attempted or at least planned.

In each case you will need to decide
- who or what to sample (how the sample is to be found).
- what questions need asking or what parameters need measuring.

Remember to analyse the results, explaining why you chose to calculate the median, mean or mode; why you chose to present the results in any particular way, and so on.

a) What is an average student?

b) Old people are more superstitious than young people.

c) More babies are born in winter than in summer.

d) Do tall people weigh more than short people?

e) Estimate the number of blades of grass on, for example, the school playing field, a football pitch or your lawn at home.

f) Any ideas suggested by any of the questions in this chapter.

Bias

If each member of the population does not have an equal chance of being selected for a sample then the sample is said to be **biased**. A biased sample will not be representative of the population.

Bias can come from a variety of sources.

- When a sample is not chosen randomly.
- When people do not reply to a questionnaire. Bias is introduced into the results as it may mean that only those who have time for filling in the form or those who wish to influence the result of the investigation have replied.
- When a replacement item is included. If you were interviewing every tenth householder in a street and one was not at home, it would be wrong to interview another householder instead. The characteristics of the person at home may be different from the one who was out.
- How, where and when data are collected. Using the internet excludes those without a computer; questioning people at a particular location excludes those who do not go there; the time of day may mean people are excluded who are not in the vicinity at that time.
- When the questions posed are not clear or are leading questions.

When you take part in a statistical exercise you must show what measures you have taken to avoid bias in the sample you have chosen.

EXERCISE 14.5

1 In each of the following, decide whether or not the method of sampling is appropriate. If it is not satisfactory, say why not.
   a) A newspaper editor wishes to gauge the public's reaction to a matter discussed in Parliament.
       She asks readers of her newspaper to write in and express their views.
   b) To find out drivers' opinions of a car park, the drivers of 50 cars in the car park are asked to complete a questionnaire.
   c) A councillor wishes to know the opinion of householders on a certain local issue. His plan is to visit every tenth house in each street. If there is no reply he has decided to call at the house next door.
   d) A supermarket thinks that as many men as women shop at their store.
       To investigate this they set up a survey one Saturday morning between 9 a.m. and 10 a.m. where they count the number of men and women going in to the store.

2 A sample of 1000 people is to be interviewed about what books they read. Comment on the following methods for obtaining the sample.
   a) Choose 1000 names from the telephone directory.
   b) Ask 1000 people at random at a local railway station one evening.
   c) Ask 100 librarians each to supply the names of 10 people.

STAGE

9

K KEY IDEAS

You should be able to

- design and write questionnaires.

- decide an appropriate sample size and how to select this sample.

You should understand

- random sampling and the use of random number tables or random number generators.

- stratified or quota sampling.

- how to avoid bias when selecting a sample.

Revision exercise D1

1   Draw a histogram to show the following distribution of the weights, to the nearest kilogram, of 50 Year 9 students.

Weight (W kg)	Frequency (f)
$32 \leqslant W < 34$	1
$34 \leqslant W < 39$	5
$39 \leqslant W < 43$	7
$43 \leqslant W < 47$	8
$47 \leqslant W < 51$	14
$51 \leqslant W < 59$	9
$59 \leqslant W < 70$	6

2   This histogram shows the masses, in grams, of plums picked in an orchard.
How many plums were picked?

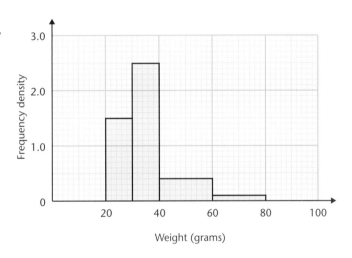

3   The heights of students in two classes are measured.
The results are given in the tables.

Class 11A	
Height (H cm)	Frequency
$130 \leqslant H < 140$	1
$140 \leqslant H < 150$	4
$150 \leqslant H < 160$	9
$160 \leqslant H < 170$	8
$170 \leqslant H < 180$	2
$180 \leqslant H < 190$	2

Class 11B	
Height (H cm)	Frequency
$120 \leqslant H < 130$	4
$130 \leqslant H < 140$	5
$140 \leqslant H < 150$	8
$150 \leqslant H < 160$	3
$160 \leqslant H < 170$	3
$170 \leqslant H < 180$	1

Show the data on two histograms.

4 In the following questions O is the centre of each circle.
In each question calculate the size of each angle or length marked with a letter.

a)

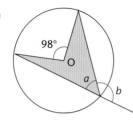

b)

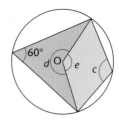

c)

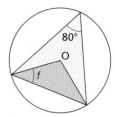

d)

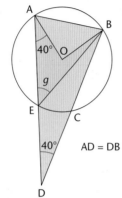

e)

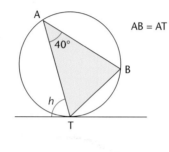

f)

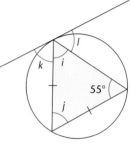

g)

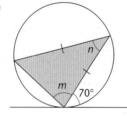

h)

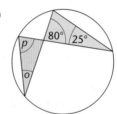

i)

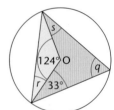

j)

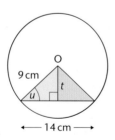

5 Find the equation of each of these lines.

a)

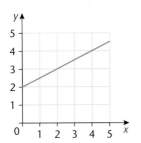

b)

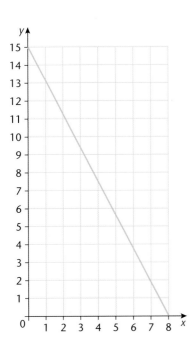

c)

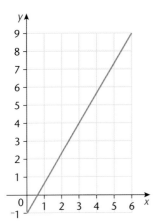

6 Find the equation of each of these lines.
   a) A line with gradient 3, passing through $(0, 2)$
   b) A line with gradient $^-2$, passing through $(1, 4)$
   c) A line with gradient $\frac{1}{2}$, passing through $(2, 6)$

7 Find the equation of the line that joins each of these pairs of points.
   a) $(4, 0)$ and $(6, 5)$
   b) $(1, 2)$ and $(6, 4)$
   c) $(2, 3)$ and $(5, ^-6)$

8 Find the equation of each of the three sides of the triangle ABC.

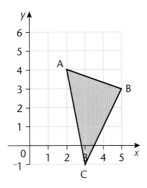

9 Find the equation of each of these lines.
   a) A line parallel to $y = 7x + 4$, passing through $(5, 1)$
   b) A line parallel to $2y = 5x + 4$, passing through $(4, 0)$
   c) A line parallel to $5x + 4y = 3$, passing through $(2, ^-1)$

10 Find the equation of each of these lines.
   a) A line perpendicular to $y = 2x + 4$, passing through $(2, 1)$
   b) A line perpendicular to $3y = 2x + 7$, passing through $(0, 1)$
   c) A line perpendicular to $2x + 5y = 4$, passing through $(1, ^-3)$

11 A mobile phone company calls 200 people, chosen at random, who subscribe to their company, to find out how satisfied they are with the service they receive.
   Is this a satisfactory method of sampling? Give a reason for your answer.

STAGE
9

12 Identify which type of sampling has been used in each of the following cases.

a) In order to determine whether the library facilities in a town are satisfactory, all the library cards are numbered and 100 questionnaires are sent out to the owners of cards selected using random numbers.

b) A factory employs 1500 people on machines, 400 on packing and distribution and 300 in the offices. A sample is taken containing 15 machine operators, 4 people from packing and distribution and 3 office workers.

13 Safia surveyed students in her school to find out their views about background music in shops.

The size of each year group in the school is shown in the table.

Year group	Boys	Girls
7	84	66
8	71	85
9	82	86
10	93	107
11	81	90
Total	411	434

Safia took a sample of 80 students.

a) Should she have sampled equal numbers of boys and girls in Year 7? Give a reason for your answer.

b) Calculate the number of students she should sample in Year 7.

14 A government officer is investigating whether there is a link between the income of families and the number of children they have.

The table shows the information in her database of families.

Number of children	Number of families
0	123
1	179
2	457
3	88
4 and over	45

She plans to interview a sample of 60 families.

How many should she select from each group to make the sample representative?

15 Sweet-tasting apples are used to make apple juice.

An apple grower needs to find out how 'sweet' a crop of apples from an orchard is.

He is advised to select a sample of 50 apples.

The orchard consists of 1500 trees, and each tree produces about 50 apples. The grower decides to pick 50 apples from one tree that he selects at random. Is this a satisfactory method? Give reasons for your answer.

16 The manager of a newly-opened gym wants to find out how successful it is. He decides to interview a sample of 50 of the members of the gym. Describe how he might choose the 50 members using each of these methods.

a) Simple random sampling

b) Systematic random sampling

c) Stratified random sampling

Stage 10 Contents

STAGE
10

STAGE
10

Using graphs to solve equations

1

You will learn about

- Solving simultaneous equations where one equation is quadratic
- Solving problems involving the intersection of a straight line with a curve
- The equation of a circle, $x^2 + y^2 = r^2$

You should already know

- How to use your calculator with trigonometrical functions and exponential functions
- Pythagoras' theorem
- How to plot the graphs of linear, quadratic, cubic and reciprocal functions
- How to solve a pair of simultaneous linear equations graphically
- How to solve a quadratic equation graphically

Quadratics

In Stage 8 you saw how you could solve a pair of simultaneous linear equations graphically. You can also use the method when one of the equations is a quadratic.

STAGE
10

EXAMPLE 1

Solve these simultaneous equations graphically.

$y = x^2 + 3x - 7$
$y = x - 4$

Use values of x from $^-5$ to $^+2$.

Make tables of values and plot the curve and the line on the same axes.

$y = x^2 + 3x - 7$

x	⁻5	⁻4	⁻3	⁻2	⁻1	0	1	2
x²	25	16	9	4	1	0	1	4
+ 3x	⁻15	⁻12	⁻9	⁻6	⁻3	0	3	6
− 7	⁻7	⁻7	⁻7	⁻7	⁻7	⁻7	⁻7	⁻7
y = x² + 3x − 7	3	⁻3	⁻7	⁻9	⁻9	⁻7	⁻3	3

$y = x - 4$

x	⁻5	0	2
y	⁻9	⁻4	⁻2

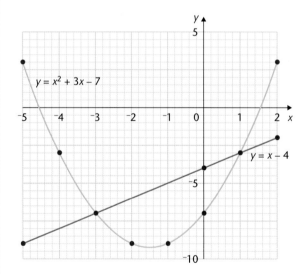

The curve and the line cross at $x = 1$, $y = ^-3$ and $x = ^-3$, $y = ^-7$.

So the solution of the simultaneous equations is $x = 1$, $y = ^-3$ or $x = ^-3$, $y = ^-7$.

EXERCISE 1.1

1 **a)** Draw the graph of $y = x^2 - 5x + 5$ for $x = {}^-2$ to ${}^+5$.

  b) On the same grid draw the graph of $2x + y = 9$.

  c) Write down the coordinates of the points where the curve and line cross.

2 **a)** Draw the graph of $y = x^2 - 3x - 1$ for $x = {}^-4$ to ${}^+3$.

  b) On the same grid draw the graph of $4x + y = 5$.

  c) Write down the coordinates of the points where the curve and line cross.

3 **a)** Draw the graph of $y = x^2 + 3$ for $x = {}^-2$ to ${}^+5$.

  b) On the same grid draw the graph of $y = 3x + 7$.

  c) Write down the coordinates of the points where the curve and line cross.

4 **a)** Draw the graph of $y = x^2 - 5x + 3$ for $x = {}^-2$ to ${}^+4$.

  b) On the same grid draw the graph of $7x + 2y = 11$.

  c) Write down the coordinates of the points where the curve and line cross.

5 Solve these simultaneous equations graphically.

$$y = 3 - x$$
$$y = 2x^2$$

6 Solve these simultaneous equations graphically.

$$x + 2y - 4 = 0$$
$$y = x^2 - 4$$

Using graphs to solve harder equations

In Stage 8, you saw how to solve equations like $x^2 - 5x + 3 = 0$ by drawing the graph of $y = x^2 - 5x + 3$ and reading off where it crossed the x-axis, where $y = 0$.

This can now be extended to include intersections with lines other than $y = 0$.

EXAMPLE 2

The graph of $y = x^2 - 3x + 1$ is shown.

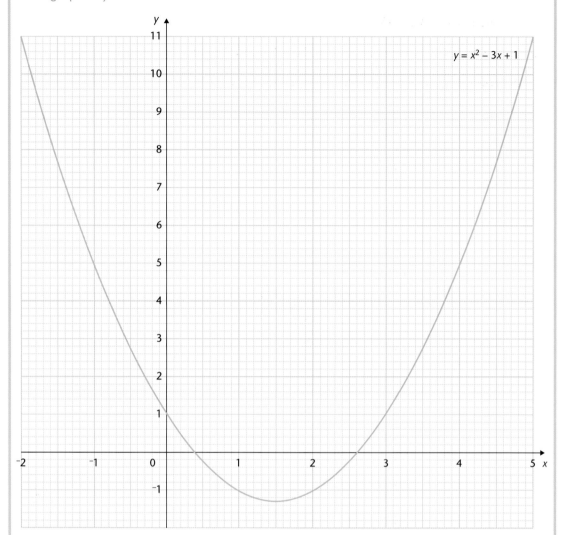

$y = x^2 - 3x + 1$

How can you use the graph to find the solution of

a) $x^2 - 3x + 1 = 0$?

b) $x^2 - 3x - 1 = 0$?

EXAMPLE 2 continued

Using graphs to solve equations

a) The solution is where the curve crosses the line $y = 0$ (the x-axis).

Solution: $x = 0{\cdot}4$ or $2{\cdot}6$.

b) $x^2 - 3x - 1 = 0$ is the same as $x^2 - 3x + 1 - 2 = 0$ or $x^2 - 3x + 1 = 2$.

Where $y = 2$ meets $y = x^2 - 3x + 1$ will give the solution.

Solution: $x = {}^-0{\cdot}3$ or $3{\cdot}3$.

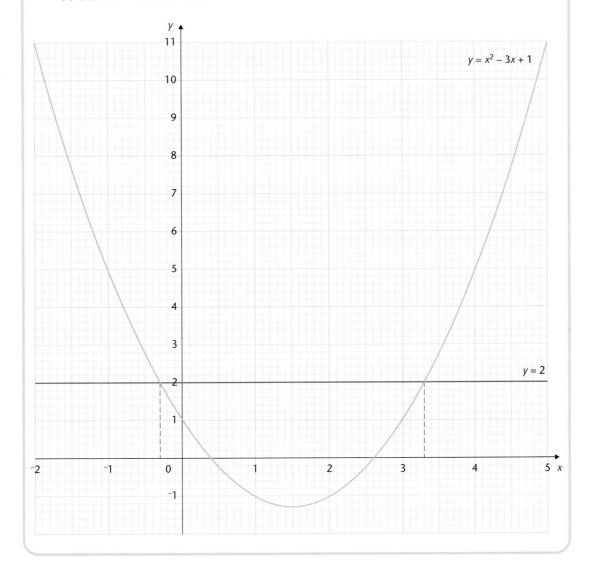

Sometimes you may have drawn a quadratic graph and then need to solve an equation that is different from the graph you have drawn. Rather than drawing another graph it may be possible to rearrange the equation to obtain the one you have drawn.

STAGE
10

EXAMPLE 3

a) Draw the graph of $y = x^2 - 1$ for $y = {}^-3$ to ${}^+3$.

b) Use your graph to solve the equation $x^2 - 2 = 0$.

c) (i) Draw a suitable line so you can find the solution of $x^2 - x - 1 = 0$.

  (ii) Solve $x^2 - x - 1 = 0$ from your graph.

a)

x	-3	-2	-1	0	1	2	3
x^2	9	4	1	0	1	4	9
– 1	-1	-1	-1	-1	-1	-1	-1
$y = x^2 - 1$	8	3	0	-1	0	3	8

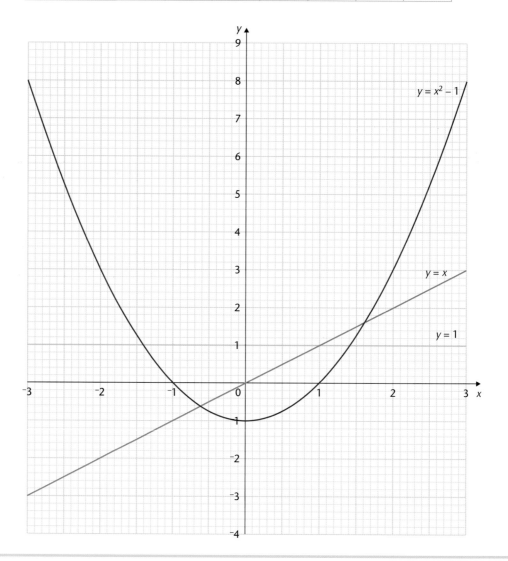

EXAMPLE 3 continued

b) $x^2 - 2 = 0$ is the same as $x^2 - 1 - 1 = 0$ or $x^2 - 1 = 1$.

Read where $y = 1$ meets $y = x^2 - 1$.

Solution: $x = {}^-1 \cdot 4$ or $1 \cdot 4$.

c) (i) $x^2 - x - 1 = 0$ is the same as $x^2 - 1 = x$. So the required line is $y = x$.

(ii) Solution: $x = {}^-0 \cdot 6$ or $1 \cdot 6$.

EXAMPLE 4

The graph of $y = x^3 - 5x$ is drawn.
(Do not draw the graph.)

What line needs to be drawn to find the solution of

$x^3 - 6x - 5 = 0$?

$x^3 - 6x - 5 = 0$ is the same as

$x^3 - 5x - x - 5 = 0$ or $x^3 - 5x = x + 5$.

The line that needs to be drawn is $y = x + 5$.

EXAM TIP

When you are asked to solve an equation from your graph, make it clear what lines you are using to find the solution. Marks will usually be given for using the correct method even if the solution is not accurate.

STAGE
10

EXAMPLE 5

a) Draw the graph of $y = x^2 - 5x + 3$ for $x = {}^-1$ to ${}^+6$.

b) From your graph solve the inequality $x^2 - 5x + 3 < 0$.

a)

x	-1	0	1	2	3	4	5	6
x^2	1	0	1	4	9	16	25	36
$- 5x$	5	0	-5	-10	-15	-20	-25	-30
$+ 3$	3	3	3	3	3	3	3	3
$y = x^2 - 5x + 3$	9	3	-1	-3	-3	-1	3	9

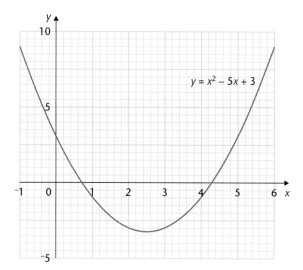

$y = x^2 - 5x + 3$

EXAM TIP

Answers should normally be given to 1 decimal place unless there is a different instruction.

b) The solution is where the curve goes below the x-axis since $y < 0$, giving $0.7 < x < 4.3$.

EXERCISE 1.2

1 a) Draw the graph of $y = x^2 - 5$ for $x = {}^-3$ to ${}^+3$.

   b) From your graph solve each of these equations.
   (i) $x^2 - 5 = 0$
   (ii) $x^2 - 3 = 0$

2 a) Draw the graph of $y = 2x^2 - 3x$ for $x = {}^-2$ to ${}^+4$.

   b) Use your graph to solve the equation $2x^2 - 3x - 5 = 0$.

3 a) Draw the graphs of $y = x^2 + 5$ and $y = 3x + 7$ for $x = {}^-2$ to ${}^+4$.

   b) What is the equation of the points where they intersect?

   c) Solve this equation from your graph.

4 a) Draw the graph of $y = 2x^2 - 10$ for $x = {}^-3$ to ${}^+3$.

   b) From your graph solve each of these equations.
   (i) $2x^2 - 10 = 0$
   (ii) $2x^2 - 3 = 0$

5 **a)** Draw the graphs of $y = x^2 + 2$ and $y = 2x + 7$ for $x = {}^-2$ to ${}^+4$.
   b) What is the equation of the points where they intersect?
   c) Solve this equation from your graph.

6 **a)** Draw the graph of $y = x^2 - 5x$ for $x = {}^-2$ to ${}^+6$.
   b) **(i)** Draw another line so that the equation of the points of intersection is $x^2 - 3x - 3 = 0$.
      (ii) Solve this equation from the graph.

7 **a)** Draw the graph of $y = x^3 - 3x$ for $x = {}^-3$ to ${}^+3$.
   b) **(i)** Draw another graph so that the equation of the points of intersection is $x^3 - 6x + 5 = 0$.
      (ii) Use the graph to solve the equation.

8 **a)** Draw the graph of $y = x^3 - 5x$ for $x = {}^-3$ to ${}^+3$.
   b) **(i)** Draw another graph so that the equation of the points of intersection is $x^3 - x^2 - 5x + 5 = 0$.
      (ii) Use the graph to solve the equation.

9 **a)** Draw the graph of $x^2 - 2x - 4$ for $x = {}^-2$ to ${}^+4$.
   b) Use your graph to find the solution of each of these.
      (i) $x^2 - 2x - 7 = 0$
      (ii) $x^2 - 4x - 6 = 0$
      (iii) $x^2 - 2x - 4 > 0$

10 **a)** Draw the graph of $x^2 + 5x + 4$ for $x = {}^-6$ to ${}^+1$.
   b) Use your graph to find the solution of each of these.
      (i) $x^2 + 5x + 4 = 0$
      (ii) $x^2 + 3x - 3 = 0$
      (iii) $x^2 + 5x + 2 < 0$

Do not draw the graphs in questions **11 to 20**.

11 The graph of $y = x^2 - 6x + 4$ is drawn. How can you use the graph to find the solutions of these equations?
   a) $x^2 - 6x + 4 = 0$
   b) $x^2 - 6x + 2 = 0$

12 The graph of $y = 2x^2 - 3x$ is drawn. How can you find the solution of $2x^2 - 3x - 5 = 0$?

13 The graph of $y = x^2 - 8x + 2$ is drawn. How can you use the graph to find the solutions of these equations?
   a) $x^2 - 8x + 2 = 0$
   b) $x^2 - 8x + 6 = 0$

14 The graph of $y = 2x^2 - x - 2$ is drawn. How can you find the solution of $2x^2 - x - 5 = 0$?

15 The graphs of $y = x^2$ and $y = 4x - 3$ are drawn on the same grid. What is the equation whose solution is found at the intersection of the two graphs?

16 The graph of $y = x^3 - 4x$ is drawn. What other graph needs to be drawn to find the solution of $x^3 - x^2 - 4x + 3 = 0$ where they cross?

17 The graphs of $y = x^2 + 3x$ and $y = 4x - 3$ are drawn on the same grid. What is the equation whose solution is found at the intersection of the two graphs?

18 The graph of $y = x^3 - 2x^2$ is drawn. What other graph needs to be drawn to find the solution of $x^3 - x^2 - 4x + 3 = 0$ where they cross?

19 The intersection of two graphs is the solution to the equation $x^2 - 4x - 2 = 0$. One of the graphs is $y = x^2 - 5x + 1$. What is the other graph?

20 The intersection of two graphs is the solution to the equation $x^2 - 5x - 3 = 0$. One of the graphs is $y = x^2 - x + 1$. What is the other graph?

Using graphs to solve equations

STAGE
10

163

The equation of a circle

Look at this equation.

$$x^2 + y^2 = 25$$

Remember Pythagoras' theorem. It looks like this equation.
Can you spot a pair of values for x and y which fit?
What about $(3, 4)$, $(4, 3)$ or $(5, 0)$? These all satisfy the equation.

So do $(^-5, 0)$, $(^-4, 3)$, $(^-3, 4)$, $(0, 5)$, $(4, ^-3)$, $(3, ^-4)$, $(0, ^-5)$, $(^-3, ^-4)$ and $(^-4, ^-3)$.

This is the graph of $x^2 + y^2 = 25$.

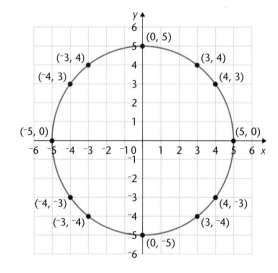

Can you see why the graph is a circle?

> **The general equation of a circle with centre (0, 0) and radius r is $x^2 + y^2 = r^2$,**
> **as shown by Pythagoras' theorem.**

▌▌ EXAMPLE 6

Draw the graph of $x^2 + y^2 = 9$.

$x^2 + y^2 = 9$ is the equation of the circle with centre (0, 0) and radius 3.

The only integer values it passes through are (⁻3, 0) (0, 3), (3, 0) and (0, ⁻3).

When $x = 2$ or $⁻2$, $4 + y^2 = 9$
$$y^2 = 5$$
$$y = \pm \sqrt{5}$$
$$= 2 \cdot 24 \text{ or } ⁻2 \cdot 24.$$

All these 8 points can be plotted and the circle will pass through them.

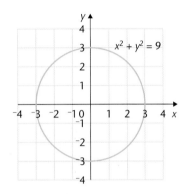

EXAM TIP

Working out the values in the usual way may cause problems!

It is easier to find the radius and use compasses rather than to calculate points.

▌▌ EXAMPLE 7

a) On the same grid draw the graphs of $x^2 + y^2 = 16$ and $y = x + 2$.

b) Use the graph to solve the two equations simultaneously, giving the answers correct to 1 decimal place.

a) $x^2 + y^2 = 16$ is the equation of the circle with centre (0, 0) and radius 4.

b) The solution is where the two graphs cross, (⁻3·6, ⁻1·6) or (1·6, 3·6).

So the solution is $x = ⁻3 \cdot 6$, $y = ⁻1 \cdot 6$ or $x = 1 \cdot 6$, $y = 3 \cdot 6$.

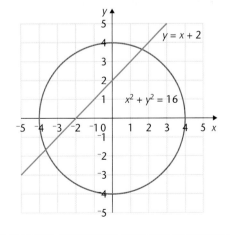

STAGE
10

EXERCISE 1.3

1 Draw the graph of $x^2 + y^2 = 64$.
Use a scale of 1 cm to 2 units for both x and y.

2 Draw the graph of $x^2 + y^2 = 625$.
Use a scale of 1 cm to 5 units for both x and y.

3 **a)** Draw the graphs of $x^2 + y^2 = 100$ and $y = x + 2$ on the same grid.
Use a scale of 1 cm to 1 units for both x and y.
  b) Find the coordinates of the points where the two graphs cross.

4 **a)** Draw the graphs of $x^2 + y^2 = 25$ and $y + x = 7$ on the same grid.
Use a scale of 1 cm to 1 unit for both x and y.
  b) Find the coordinates of the points where the two graphs meet.

5 **a)** Draw the graphs of $x^2 + y^2 = 64$ and $y = 2x + 8$ on the same grid.
Use a scale of 1 cm to 2 units for both x and y.
  b) Use the graph to solve the simultaneous equations $x^2 + y^2 = 64$ and $y = 2x + 8$.
Give the answers either as whole numbers or correct to 1 decimal place.

6 **a)** Draw the graphs of $x^2 + y^2 = 49$ and $y = x + 5$ on the same grid.
Use a scale of 1 cm to 2 units for both x and y.
  b) Use the graph to solve the simultaneous equations $x^2 + y^2 = 49$ and $y = x + 5$.
Give the answers either as whole numbers or correct to 1 decimal place.

K KEY IDEAS

- When solving simultaneous equations graphically, draw the two graphs and find where they intersect.

- Sometimes, when solving an equation graphically, you will need to adapt the equation. Rearrange it so that the equation of the graph drawn is on the left of the equals sign. What appears on the right is the equation of the second graph that needs to be drawn.

- The equation of a circle with centre (0, 0) and radius r is $x^2 + y^2 = r^2$.

STAGE
10

Growth and decay

Exponential growth

£200 is invested at 5% per year compound interest.

After 1 year the investment will be worth $200 \times 1 \cdot 05 = £210$.
After 2 years the investment will be worth $210 \times 1 \cdot 05 = £220 \cdot 50$.
After 3 years the investment will be worth $220 \cdot 50 \times 1 \cdot 05 = £231 \cdot 53$. And so on.

This sort of calculation, involving a constant multiplier, is an example of **exponential growth**. If the multiplier is a large number, the figures get very large, very quickly.

To do the calculation in the same way for a large number of years is time consuming. It would be easier, if it was possible, to find a formula for the calculation.

Look again at the calculations for 2 and 3 years.

Another way of looking at the calculation for 2 years is
$200 \times 1 \cdot 05 \times 1 \cdot 05 = 200 \times 1 \cdot 05^2$.
Similarly, the calculation for 3 years is
$200 \times 1 \cdot 05 \times 1 \cdot 05 \times 1 \cdot 05 = 200 \times 1 \cdot 05^3$.

So, after 20 years, the investment will be worth $£200 \times 1 \cdot 05^{20}$.

On a calculator this calculation is done using the powers key. This is usually labelled $\boxed{\wedge}$.

The calculation is then simply $200 \times 1 \cdot 05$ $\boxed{\wedge}$ $20 = £530 \cdot 66$.

The formula for this calculation is $A = 200 \times 1 \cdot 05^n$, where £$A$ is the amount the investment is worth and n is the number of years.

2

Growth and decay

A | ACTIVITY 1

DOUBLE YOUR MONEY

Find out how long it would be before an investor's money doubled.

Use your calculator to calculate how long it would take if the interest rate was 5%.

Then use a spreadsheet and compare the times for different rates of interest.

▌▌ EXAMPLE 1

The number of bacteria present in a population doubles every hour.

If there are 500 present at 12 noon, find the number present

a) at 2 p.m.

b) at 3 p.m.

c) at midnight.

d) after n hours.

a) $500 \times 2 \times 2 = 2000$

b) $500 \times 2^3 = 4000$ or $2000 \times 2 = 4000$

c) $500 \times 2^{12} = 2\,048\,000$

d) 500×2^n

STAGE
10

Exponential decay

A car depreciates in value by 15% per year. It cost £12000 when it was new.

After 1 year it will be worth $12\,000 \times 0{\cdot}85 = £10\,200$.

After 2 years it will be worth $10\,200 \times 0{\cdot}85 = £8670$.

After 3 years it will be worth $8670 \times 0{\cdot}85 = £7369{\cdot}50$. And so on.

This calculation, where the constant multiplier is less than 1, is an example of **exponential decay**.

The calculations involved work exactly like the ones for exponential growth, except that the multiplier is less than 1 and so the numbers get smaller instead of bigger.

After 10 years the car will be worth $12\,000 \times 0{\cdot}85^{10} = £2362{\cdot}49$.

The formula for this calculation is $A = 12\,000 \times 0{\cdot}85^{n}$,

where £A is the amount the car is worth and n is the number of years.

EXAMPLE 2

The population of a certain species of bird is dropping by 20% every 10 years. If there were 50000 in 1970, how many will there be

a) in 2010?

b) in 2020?

c) in 2100?

d) n years after 1970?

a) $\dfrac{2010 - 1970}{10} = 4$ $50\,000 \times 0{\cdot}8^4 = 20\,480$

b) $20\,480 \times 0{\cdot}8 = 16\,384$ (or $50\,000 \times 0{\cdot}8^5 = 16\,384$)

c) $\dfrac{2100 - 1970}{10} = 13$ $50\,000 \times 0{\cdot}8^{13} = 2749$

d) Since the population decreases by 20% every 10 years, the number will be $50\,000 \times 0{\cdot}8^{\frac{n}{10}}$.

STAGE 10

The calculations in Examples 1 and 2 show the dramatic changes that exponential growth and decay can make.

Exponential decay can also be achieved by using a **negative power**.

For example, if a population starts at 1 million and is halved every year,

after 5 years the population is $1\,000\,000 \times 0{\cdot}5^5 = 31\,250$

after n years the population is $1\,000\,000 \times 0{\cdot}5^n$.

Since $0{\cdot}5 = \frac{1}{2} = 2^{-1}$, the calculations can be written as

after 5 years the population is $1\,000\,000 \times 2^{-5} = 31\,250$

after n years the population is $1\,000\,000 \times 2^{-n}$.

Graphs of exponential growth and decay functions

In Example 1 the number of bacteria present doubled every hour. To show this graphically, first make a table showing the number of bacteria in the populaton over a period of time, say 4 hours.

Number of hours (n)	0	1	2	3	4
Number of bacteria (B)	500	1000	2000	4000	8000

Then plot the points on a graph.

The shape of this graph is typical of an exponential growth function.

Exponential growth functions have an equation of the form $y = ab^x$.

The equation of this graph is $B = 500 \times 2^n$.

■ 500 because this is the initial value.
■ 2 because the number of bacteria doubles each time.

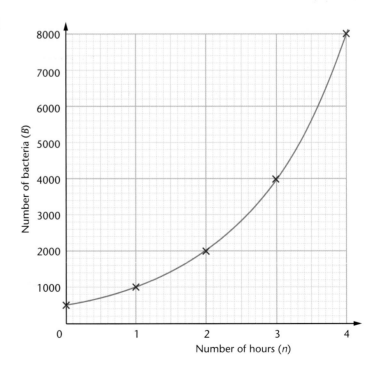

Similarly, you can show the information in Example 2 graphically.

In Example 2 the population of a certain species of bird dropped by 20% every 10 years.

First make a table showing the population over 50 years.

Year (n)	1970	1980	1990	2000	2010	2020
Number of birds (B)	50 000	40 000	32 000	25 600	20 480	16 384

Then plot the points on a graph.

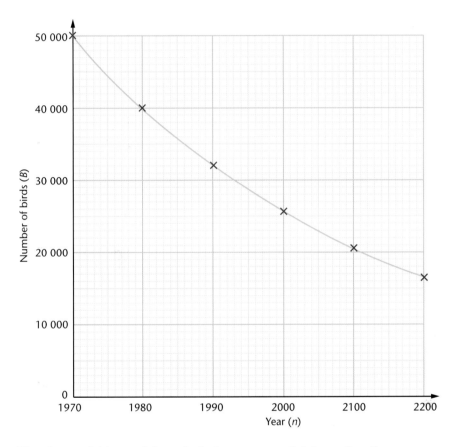

The shape of this graph is typical of an exponential decay function.

The equation of this graph is $B = 50\,000 \times 0.8^{\frac{n}{10}}$.

■ 50 000 because this the initial value.
■ 0·8 because the population is decreasing by 20% (0·2 each time).
■ n is divided by 10 because the increase is every 10 years, not every year.

In the same way as you used graphs to solve quadratic equations in Chapter 1 and earlier in the course, you can use graphs to solve exponential equations.

EXAMPLE 3

Plot a graph of $y = 3^x$ for values of x from $^-2$ to 3.

Use your graph to estimate
a) the value of y when $x = 2\cdot4$.
b) the solution to the equation $3^x = 20$.

Make a table of values and plot the graph.

x	-2	-1	0	1	2	3
y	0·111	0·333	1	3	9	27

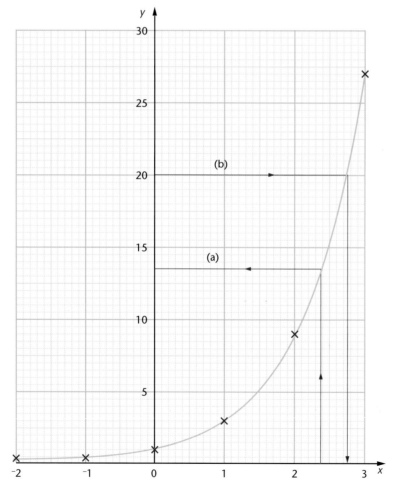

a) Draw the line $x = 2\cdot4$ and find where it intersects the curve. $y = 13\cdot5$

b) Draw the line $y = 20$ and find where it intersects the curve. $x = 2\cdot75$

EXAMPLE 4

Plot a graph of $y = 2^{-x}$ for values of x from $^-4$ to 2.

Use your graph to estimate

a) the value of y when $x = 0.5$.

b) the solution to the equation $2^{-x} = 10$.

Make a table of values and plot the graph.

x	$^-4$	$^-3$	$^-2$	$^-1$	0	1	2
y	16	8	4	2	1	0·5	0·25

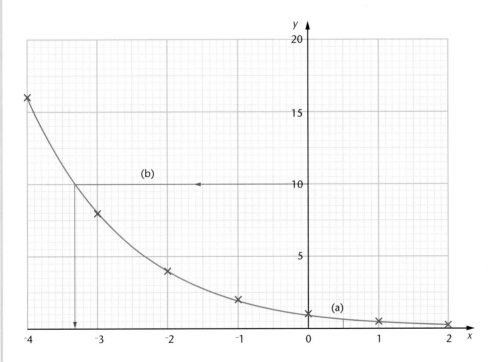

a) Draw the line $x = 0.5$ and find where it intersects the curve.

$y = 0.7$

b) Draw the line $y = 10$ and find where it intersects the curve.

$x = ^-3.35$

Note that, although the scales make it difficult to show here, neither graph ever reaches the x-axis.

EXERCISE 2.1

1. Two people start a rumour that goes round a school.
The following table shows the number of people (n) who have heard the rumour after t hours.

Time (t) in hours	0	1	2	3	4	5
Number of people (n)	2	6	18	54	162	486

Find a formula for n in terms of t.

2. The mass (m) of a chemical present t minutes after the start of a chemical reaction is given in the table below.

Time (t) in minutes	0	1	2	3	4	5
Mass (m) in grams	100	50	25	12·5	6·25	3·125

Find a formula for m in terms of t.

3. Plot a graph of $y = 2^x$ for values of x from -2 to 5.
Use a scale of 2 cm to 1 unit on the x-axis and 2 cm to 5 units on the y-axis.
Use your graph to estimate
   a) the value of y when $x = 3\cdot2$.
   b) the solution to the equation $2^x = 20$.

4. Plot a graph of $y = 1\cdot5^x$ for values of x from -3 to 5.
Use a scale of 2 cm to 1 unit on both axes.
Use your graph to estimate
   a) the value of y when $x = 2\cdot4$.
   b) the solution to the equation $1\cdot5^x = 6$.

5. Copy and complete the table of values for $y = 3^{-x}$.

x	0	0·5	1	1·5	2	2·5	3	3·5	4
y	1								

Plot the graph of $y = 3^{-x}$ for these values.
Use a scale of 2 cm to 1 unit on the x-axis and 1 cm to 0·1 unit on the y-axis.
Use your graph to estimate
   a) the value of y when $x = 1\cdot2$.
   b) the solution to the equation $3^{-x} = 0\cdot1$.

EXERCISE 2.1 continued

6 Copy and complete the table of values for $y = 4^{-x}$.

x	⁻2·5	⁻2	⁻1·5	⁻1	⁻0·5	0	0·5	1
y		16						

Plot the graph of $y = 4^{-x}$ for these values.
Use a scale of 2 cm to 1 unit on the *x*-axis and 2 cm to 5 units on the *y*-axis.
Use your graph to estimate
a) the value of *y* when $x = {}^{-}1{\cdot}8$.
b) the solution to the equation $4^{-x} = 25$.

Using trial and improvement to solve growth and decay problems

Look again at the graph on page 170 showing the number of bacteria in the population in Example 1. As you have seen, you can use the graph to find the answer to questions such as 'How many hours after 12 noon is the population 6000?'.

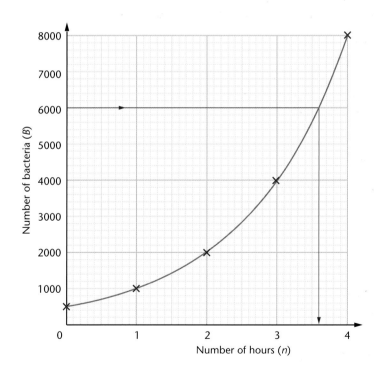

From the graph, you can see that the answer is approximately 3·6.

STAGE
10

Whenever you are using a graph, however, your answer will only be approximate and, because of the shape of exponential graphs, the range of values on the y-axis is often large, making it difficult to obtain a good estimate.

You can obtain a more accurate estimate using **trial and improvement**. This requires the use of the powers key on your calculator. The solution you obtained graphically gives you the first value to test.

The following table demonstrates a possible way of obtaining the solution correct to 2 decimal places.

Number of hours (n)	Number of bacteria = 500×2^n	Decision
3·6	6062·86…	Too large.
3·5	5656·85…	Too small but further away. Try between 3·5 and 3·6 but near 3·6.
3·59	6020·987…	Too large, so try smaller.
3·58	5979·397…	Too small, so try halfway between.
3·585	6000·156…	Too large. Solution is between 3·58 and 3·585, so it is 3·58 correct to 2 decimal places.

EXAMPLE 5

Solve $2^{-x} = 0.6$.

Use trial and improvement and give your solution correct to 1 decimal place.

x	2^{-x}	Decision
0	1	Too large.
1	0·5	Solution is between 0 and 1. Try halfway between.
0·5	0·707…	Solution is between 0·5 and 1. Try 0·7.
0·7	0·615…	Solution is between 0·7 and 1. Try 0·8.
0·8	0·574…	Solution is between 0·7 and 0·8. Try 0·75.
0·75	0·594…	Solution is between 0·7 and 0·75, so it is 0·7 to 1 d.p.

EXERCISE 2.2

1 a) Copy and complete the following table for a sum of money £y growing at 12% each year for x years.

x	y
0	500
1	560
2	627·20
3	
4	
5	

b) Use trial and improvement to solve the equation $500 \times 1·12^x = 750$, giving the value of x correct to 1 decimal place.

2 a) Copy and complete this table for the function $y = 3^x$.

x	y
0	
1	
2	
3	
4	
5	

b) Find the value of y when $x = 12$.
c) Use trial and improvement to find the value of x, correct to 1 decimal place, when $y = 6000$.

3 a) Sketch the graph of $y = 2^x$ for $x = 0$ to 6.
b) Use trial and improvement to find the value of x, correct to 1 decimal place, when $y = 200$.

4 The number of bacteria in a population doubles every 15 minutes. At the start of measuring time, there were 50 bacteria.

a) Copy and complete the table to show the number of bacteria during the first two hours.

Time (hours)	Number of bacteria
0·00	50
0·25	
0·50	
0·75	
1·00	
1·25	
1·50	
1·75	
2·00	

b) Calculate the population after 6 hours. Give your answer in standard form, to 2 significant figures.
c) Find how long it took for the population to reach 4000. Give your answer in hours, correct to 1 decimal place.

5 The size, y, of a population of bacteria is growing according to the rule $y = 25 \times 1·02^t$, where t minutes is the measured time.

a) How many bacteria are there at time $t = 0$?
b) What will the population be 5 hours after starting to measure the time?
c) Find how long it takes the population to double in size.

STAGE
10

6 In 2000, the value of Bharat's stamp collection was £85. Assume the value increases at 5% each year.
   a) What was its value in 2004?
   b) Write, as simply as possible, an expression for its value x years after 2000.
   c) Find by calculation how many years it will take for the value of the stamp collection to double.

7 A population of bats is declining at a rate of 15% each year.

At the start of 2000 there were 140 bats. How many years after the start of 2000 had the population fallen to 80 bats? Give your answer to 1 decimal place.

8 A sample of a radioactive element has a mass of 50 g. Decay reduces its mass by 10% each year.
   Use trial and improvement to estimate the time taken for its mass to halve. Give your answer in years to 1 decimal place.

9 £5000 is invested at 3% compound interest per year.
   a) Calculate the value of the investment after 20 years.
   b) Find, to the nearest year, the time taken for the investment to be worth £12 000.

10 A car cost £16 000 when new and depreciates in value by 16% each year.
   a) State a formula for the value of the car after n years.
   b) Find, in years to 1 decimal place, the age of the car when its value is £5000.

11 The graph shows a function of the form $y = ab^x$. Find the values of a and b.

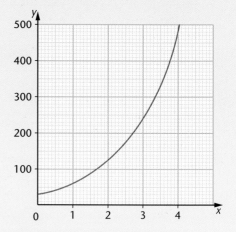

12 The value of an investment is declining at a rate of 10% per year. Initially the investment was worth £2000.
   a) Calculate how much it is worth after 4 years.
   b) Use trial and improvement to find how long after the start the investment will be worth £1000, giving your answer in years correct to 1 decimal place.

13 A population of bacteria is decreasing at a rate of 14% per hour.
   Initially it is estimated that there are 2000 bacteria.
   a) Sketch a graph to show this population in the first 6 hours.
   b) Find, correct to 1 decimal place, the number of hours after which the population is 750.

14 The value of an antique increases by 10% each year. At the beginning of 2001 it was worth £450.
   How many years later will it be worth £2000?
   Give your answer correct to 1 decimal place.

EXERCISE 2.2 continued

15 £2000 is invested at 6% compound interest.
   a) Write a formula for the value of this investment n years later.
   b) Find, to the nearest year, the time taken for the investment to be worth £6000.

16 The mass (m grams) of a chemical present t minutes after the start of a chemical reaction is given by the formula $m = 100 \times 0.5^t$.
   a) What mass was present initially?
   b) Find, correct to 2 decimal places, the value of t when the mass was 20 g.

17 Elaine buys a car for £9000. It depreciates in value by 12% each year.
   a) Write down a formula for the value, v, of the car after t years.
   b) Calculate the value of the car after
      (i) 3 years.
      (ii) 8 years.
   c) Elaine wants to sell the car when its value has reduced to £5000. Use trial and improvement to work out how many years she will keep the car.

18 The number of a strain of bacteria decays exponentially following the formula $N = 1\,000\,000 \times 2^{-t}$, where N is the number of bacteria present and t is the time in hours.
   a) How many bacteria were present originally?
   b) How many bacteria were present after
      (i) 5 hours?
      (ii) 12 hours?
   c) After how many hours will the bacteria no longer exist; that is, after how many hours will be the number of bacteria be less than 1?

19 The curve $y = ab^x$ passes through the points $(0, 5)$ and $(3, 20.48)$. Find the values of a and b.

20 Use trial and improvement to solve these equations, giving your solution correct to 2 decimal places.
   a) $2^x = 20$
   b) $3^{-x} = 0.5$
   c) $1.5^x = 6$
   d) $4^{-x} = 0.1$
   e) $6^x = 0.03$

KEY IDEAS

- Exponential growth occurs when a value is multiplied by a constant multiplier (greater than 1).

- Exponential decay occurs when a value is multiplied by a constant multiplier (less than 1).

- Exponential functions are functions of the form $y = ab^x$. Since $b^0 = 1$, the graphs of these functions always go through $(0, a)$.

- To solve an exponential equation when the power is unknown, use the powers key on your calculator and trial and improvement.

Rational and irrational numbers

You will learn about

- Rational and irrational numbers
- Using prime factors to identify fractions which represent terminating decimals
- Converting any recurring decimal to a fraction and vice versa
- Simplifying expressions involving surds
- Simplify expressions by rationalising a denominator

You should already know

- What natural numbers and integers are
- What terminating decimals and recurring decimals are
- The 'dot' notation for recurring decimals
- How to multiply out and simplify brackets like $(a + b)(c + d)$

Rational numbers

To solve equations such as $x + 3 = 8$, the only set of numbers necessary is the set of **natural numbers** 1, 2, 3, 4, 5, ...

To solve equations such as $x + 8 = 3$, it is necessary to introduce negative numbers to our set of numbers.

This gives the **integers** ... $^-4$, $^-3$, $^-2$, $^-1$, 0, 1, 2, 3, 4, ...

To solve equations such as $4x = 3$ or $3x = 8$ or $^-7x = 5$, numbers like $\frac{3}{4}$, $\frac{8}{3} = 2\frac{2}{3}$, $\frac{^-5}{7}$ ... need to be introduced.

These numbers are called **rational numbers**. A rational number is one that can be written as a fraction with integers as the numerator and the denominator.

As well as the obvious fractions, rational numbers include

- natural numbers; for example, 5 can be written as $\frac{5}{1}$
- integers; for example, $^-6$ can be written as $\frac{^-6}{1}$
- terminating decimals; for example, 3·24 can be written as $\frac{324}{100}$
- recurring decimals; for example, $0·\dot{6} = 0·666\,66...$ can be written as $\frac{2}{3}$.

Irrational numbers

The question arises as to what numbers are left?

To solve equations such as $x^2 = 2$, numbers such as $\sqrt{2}$ need to be introduced.

$\sqrt{2} = 1.414\,213\,562\ldots$

This is a decimal that neither terminates nor recurs, so it cannot be written as a fraction.

These numbers are called **irrational numbers** as they cannot be written as a fraction with two integers. They are decimals that go on forever without recurring.

Irrational numbers include

- all non-exact square roots, for example, $\sqrt{7}$ but *not* $\sqrt{16}$ or $\sqrt{\frac{9}{16}}$

- special numbers that occur in mathematics. The only one that you have met so far is π, which is $3.141\,592\,654\ldots$. Another that is on your calculator but which you will not use unless you study mathematics beyond GCSE is e, which is $2.718\,281\,828\,46\ldots$.

It can be proved that $\sqrt{2}$ is irrational but that is not necessary at this stage.

Since the rational numbers include the natural numbers and integers, and the irrational numbers are all those numbers that are not rational, these two sets comprise all the **real** numbers that you will need to use at GCSE.

▌▌ EXAMPLE 1

State which of these numbers are irrational.

a) $\frac{3}{4}$

b) $\frac{2}{3}$

c) $\sqrt{11}$ ✓ Irrational

d) 2π ✓

e) 3.142 Rational

f) $-1\frac{1}{4}$.

Clearly **a)** and **b)** are rational.

3.142, although an approximation to π, is a terminating decimal and can be written as $\frac{3142}{1000}$.

$-1\frac{1}{4}$ can be written as $\frac{-5}{4}$.

The only two of these numbers that cannot be written as fractions are **c)** and **d)**, so these are irrational.

STAGE
10

C CHALLENGE 1

Find a proof that $\sqrt{2}$ is irrational or try to prove it yourself.

There is an elegant proof by contradiction which starts like this.

Assume $\sqrt{2}$ is rational.

Then $\sqrt{2}$ is $\dfrac{a}{b}$ for some integers a and b whose only common factor is 1.

Squaring: $2 = \dfrac{a^2}{b^2} \ldots$

Hint
If a number is expressed as $2n$, what do you know about that number?

Terminating and recurring decimals

Fractions can easily be converted to decimals by dividing the numerator by the denominator. For example, $\frac{5}{8} = 5 \div 8 = 0.625$.

You met recurring decimals and terminating decimals in Stage 7.

$\frac{5}{8} = 0.625$ and is a terminating decimal but $\frac{2}{3} = 0.66666\ldots$ which is a recurring decimal.

Here are some other fractions which give terminating decimals.

$\frac{1}{5} = 0.2,\quad \frac{3}{4} = 0.75\quad$ and $\quad \frac{7}{8} = 0.875$

Here are some other fractions which give recurring decimals.

$\frac{1}{6} = 0.1666\ldots = 0.1\dot{6},\quad \frac{1}{7} = 0.142857142 = 0.\dot{1}42857\dot{7}\quad$ and $\quad \frac{4}{9} = 0.444444\ldots = 0.\dot{4}$

The fractions which give terminating decimals have denominators which only have prime factors of 2 and/or 5. These are the factors of 10.

EXAM TIP
The dots over the digits show how many recur. If it is more than two, only the first and last have a dot.

So, if the denominator of a fraction only has prime factors which are factors of 10, it will give a terminating decimal. If the denominator of a fraction has prime factors which are not factors of 10, it will give a recurring decimal.

EXAMPLE 2

State, giving your reasons, whether each of these fractions will give a terminating decimal or a recurring decimal.

a) $\frac{4}{25}$ **b)** $\frac{3}{15}$ **c)** $\frac{7}{40}$ **d)** $\frac{8}{11}$ **e)** $\frac{29}{30}$

a) $\frac{4}{25}$ terminates because $25 = 5^2$ and 5 is a factor of 10.

b) $\frac{3}{15}$ terminates because $\frac{3}{15} = \frac{1}{5}$ and 5 is a factor of 10.

c) $\frac{7}{40}$ terminates because $40 = 2^3 \times 5$ so all the prime factors of the denominator are factors of 10.

d) $\frac{8}{11}$ recurs because 11 is prime.

e) $\frac{29}{30}$ recurs because $30 = 2 \times 3 \times 5$ and 3 is not a factor of 10.

ACTIVITY 1

Investigate patterns of recurring decimals.

- In groups, without a calculator, share the work to find $\frac{1}{7}, \frac{2}{7}, \frac{3}{7}, \frac{4}{7}, \frac{5}{7}, \frac{6}{7}$ as decimals, and compare your results.

- Similarly, have a quick look at what happens with $\frac{1}{9}, \frac{2}{9}$, etc., and $\frac{1}{11}, \frac{2}{11}$, etc.

- More difficult, but more interesting, are the thirteenths and seventeenths. A calculator may be of limited help here.

Converting recurring decimals to fractions

Terminating decimals can be converted to fractions using your knowledge of place value.

$0{\cdot}1 = \frac{1}{10}$ and $0{\cdot}02 = \frac{2}{100} = \frac{1}{50}$, and similarly for other terminating decimals.

Converting recurring decimals to fractions is somewhat more difficult. It is worth remembering that $0{\cdot}\dot{5} = \frac{5}{9}$ and similarly $0{\cdot}\dot{7} = \frac{7}{9}$, etc., but if more figures recur a more formal method is needed.

This method is illustrated in the following example.

EXAMPLE 3

Express $0.\dot{4}\dot{2}$ as a fraction in its lowest terms.

Let $r = 0.\dot{4}\dot{2}$.	$r = 0.424\,242\,42...$
Multiply by 100	$100r = 42.424\,242\,42...$
Subtract	$r = \underline{0.424\,242\,42...}$
	$99r = 42$

$$r = \frac{42}{99}$$

$$= \frac{14}{33}$$

So $0.\dot{4}\dot{2} = \frac{14}{33}$

The method always works provided you multiply by the correct power of 10.

If one figure recurs, multiply by 10.
If two figures recur, multiply by 100.
If three figures recur, multiply by 1000 and so on.

EXAMPLE 4

Write $0.4\dot{2}0\dot{7}$ as a fraction in its lowest terms.

Let $r = 0.4\dot{2}0\dot{7}$	$r = 0.420\,720\,7207...$	Note: three figures recur.
Multiply by 1000	$1000r = 420.720\,720\,7207...$	
Subtract	$r = \underline{0.420\,720\,7207...}$	
	$999r = 420.3$	

$$r = \frac{420.3}{999}$$

$$= \frac{4203}{9990}$$

$$= \frac{467}{1110}$$

So $0.4\dot{2}0\dot{7} = \frac{467}{1110}$

For numbers such as $3.4\dot{2}0\dot{7}$ the 3 can be added at the end.

So $3.4\dot{2}0\dot{7} = 3\frac{467}{1100}$

EXERCISE 3.1

1 State whether each of these numbers is rational or irrational, showing how you decide.

a) $\frac{17}{20}$ b) $0\cdot46$

c) $\sqrt{\frac{2}{25}}$ d) $\sqrt{169}$

e) 5π f) $3\cdot141\,59$

g) $^-0\cdot23\dot{4}$ h) $5 + \sqrt{3}$

i) $^-6\sqrt{2}$ j) $\sqrt{\frac{4}{25}}$

k) $0\cdot49$ l) $0\cdot5\dot{3}$

m) $\sqrt{324}$ n) $\sqrt{27}$

o) $5\pi + 2$ p) $^-2\cdot718$

q) $\frac{4\pi}{3\pi}$ r) $2\sqrt{3} + \sqrt{5}$

s) $\sqrt{2} - 7$ t) $\sqrt{1\frac{7}{9}}$

2 State, giving your reasons, whether each of these fractions will terminate or recur.

a) $\frac{2}{5}$ b) $\frac{2}{17}$

c) $\frac{38}{125}$ d) $\frac{7}{18}$

e) $\frac{3}{8}$ f) $\frac{4}{15}$

g) $\frac{3}{20}$ h) $\frac{4}{35}$

i) $\frac{9}{120}$ j) $\frac{11}{12}$

3 Find the fraction equivalent of each of these terminating decimals.
Write each fraction in its simplest form.

a) $0\cdot12$ b) $0\cdot205$

c) $0\cdot375$ d) $0\cdot3125$

4 Convert these fractions to recurring decimals using the dot notation.

a) $\frac{7}{11}$ b) $\frac{3}{7}$

c) $\frac{3}{70}$ d) $\frac{23}{90}$

e) $\frac{2079}{4995}$ f) $\frac{7}{33}$

g) $\frac{5}{13}$ h) $\frac{5}{1300}$

i) $\frac{17}{36}$ j) $\frac{481}{1100}$

5 Convert these recurring decimals to fractions or mixed numbers in their lowest terms.

a) $0\cdot\dot{2}$ b) $0\cdot\dot{7}$

c) $0\cdot\dot{4}\dot{8}$ d) $0\cdot2\dot{3}$

e) $0\cdot1\dot{3}\dot{2}$ f) $0\cdot\dot{4}\dot{3}$

g) $0\cdot40\dot{2}$ h) $0\cdot2\dot{3}\dot{6}$

i) $0\cdot3\dot{5}$ j) $0\cdot5\dot{4}$

k) $0\cdot1\dot{2}$ l) $0\cdot1\dot{7}$

m) $0\cdot\dot{1}23\dot{4}$ n) $2\cdot1\dot{8}$

Simplifying surds

An expression containing an irrational number, such as $\sqrt{3}$ or $6 + 2\sqrt{5}$, is called a **surd**.

Surds can often be simplified by using the result

$$\sqrt{a \times b} = \sqrt{a} \times \sqrt{b}.$$

This result can be demonstrated using $\sqrt{36}$.

$$\sqrt{36} = 6 = 2 \times 3 = \sqrt{4} \times \sqrt{9}$$

STAGE
10

EXAMPLE 5

Simplify $\sqrt{50}$.

$\sqrt{50} = \sqrt{25 \times 2} = \sqrt{25} \times \sqrt{2} = 5\sqrt{2}$

EXAM TIP

Look for as large a factor of the number as possible which has an exact square root. Here, this is 25. In Example 6, it is 36.

EXAMPLE 6

Simplify $\sqrt{72}$.

9 is a factor of 72   so $\sqrt{72} = \sqrt{9} \times \sqrt{8} = 3\sqrt{8}$

However, 4 is a factor of 8 so $3 \times \sqrt{8} = 3 \times \sqrt{4} \times \sqrt{2}$

$$= 3 \times 2 \times \sqrt{2}$$
$$= 6\sqrt{2}$$

Or, if you spot straight away that 36 is a factor of 72, $\sqrt{72} = \sqrt{36 \times 2}$

$$= \sqrt{36} \times \sqrt{2}$$
$$= 6\sqrt{2}$$

EXAMPLE 7

Simplify $\sqrt{12} \times \sqrt{27}$.

Method 1

$\sqrt{12} = \sqrt{4 \times 3}$ $\sqrt{27} = \sqrt{9 \times 3}$

$\phantom{\sqrt{12}} = \sqrt{4} \times \sqrt{3}$ $\phantom{\sqrt{27}} = \sqrt{9} \times \sqrt{3}$

$\phantom{\sqrt{12}} = 2 \times \sqrt{3}$ $\phantom{\sqrt{27}} = 3 \times \sqrt{3}$

So $\sqrt{12} \times \sqrt{27} = 2 \times \sqrt{3} \times 3 \times \sqrt{3}$

$$= 2 \times 3 \times \sqrt{3} \times \sqrt{3}$$
$$= 6 \times 3$$
$$= 18$$

Method 2

$\sqrt{12} \times \sqrt{27} = \sqrt{12 \times 27}$

$$= \sqrt{324}$$
$$= 18$$

STAGE

10

so $\sqrt{36} = \sqrt{4 \times 9} = \sqrt{4} \times \sqrt{9}$

Note: By definition of what we mean by a square root, $\sqrt{a} \times \sqrt{a} = a$.

Although Method 2 is probably easier if you have a calculator, Method 1 is probably easier if you cannot use a calculator.

Example 7 also illustrates the fact that the product of two irrational numbers can be rational.

Manipulation of expressions of the form $a + b\sqrt{c}$

An expression such as $2 + \sqrt{3}$, which is the sum of a rational number and an irrational number, is irrational.

This is because $2 + 1{\cdot}732\,050\,808... = 3{\cdot}732\,050\,808...$ which is itself a decimal which goes on forever without recurring and so is irrational.

EXAMPLE 8

If $x = 5 + \sqrt{3}$ and $y = 3 - 2\sqrt{3}$, simplify these.

a) $x + y$

b) $x - y$

c) xy

a) $\begin{aligned} x + y &= 5 + \sqrt{3} + 3 - 2\sqrt{3} \\ &= 5 + 3 + \sqrt{3} - 2\sqrt{3} \\ &= 8 - \sqrt{3} \end{aligned}$

b) $\begin{aligned} x - y &= 5 + \sqrt{3} - (3 - 2\sqrt{3}) \\ &= 5 + \sqrt{3} - 3 + 2\sqrt{3} \\ &= 5 - 3 + \sqrt{3} + 2\sqrt{3} \\ &= 2 + 3\sqrt{3} \end{aligned}$

These two results illustrate the fact that when adding and subtracting these numbers, you can deal with the rational and irrational parts separately.

c) $\begin{aligned} xy &= (5 + \sqrt{3})(3 - 2\sqrt{3}) = 15 - 10\sqrt{3} + 3\sqrt{3} - 2\sqrt{3} \times \sqrt{3} \\ &= 15 - 2 \times 3 - 10\sqrt{3} + 3\sqrt{3} \\ &= 9 - 7\sqrt{3} \end{aligned}$

These expressions can be manipulated using the ordinary rules of algebra and arithmetic.

STAGE
10

EXAMPLE 9

If $x = 5 + \sqrt{2}$ and $y = 3 - \sqrt{2}$, simplify these.

a) $x + y$

b) y^2

a) $x + y = 5 + \sqrt{2} + 3 - \sqrt{2}$
$ = 5 + 3 + \sqrt{2} - \sqrt{2}$
$ = 8$

Note that this result indicates that it is possible for the sum of two irrational numbers to be rational.

b) $y^2 = (3 - \sqrt{2})^2$
$ = (3 - \sqrt{2})(3 - \sqrt{2})$
$ = 9 - 3\sqrt{2} - 3\sqrt{2} + \sqrt{2} \times \sqrt{2}$
$ = 9 + 2 - 6\sqrt{2}$
$ = 11 - 6\sqrt{2}$

Note that this result is also an application of the algebraic result $(a + b)^2 = a^2 + 2ab + b^2$.

Rationalising denominators

When dealing with fractions it is often preferable to have the numerator as an irrational number rather than the denominator. This is particularly so when no calculator is available as it is far easier to divide a long decimal by a whole number than to divide a whole number by a long decimal.

It is possible, using the rules of fractions, to convert numbers with irrational denominators to ones with irrational numerators. This method is illustrated in the following example.

EXAMPLE 10

Rationalise the denominator in these irrational fractions.

a) $\dfrac{5}{\sqrt{2}}$

b) $\dfrac{7}{\sqrt{12}}$

EXAMPLE 10

a) Multiply the numerator and the denominator by $\sqrt{2}$. By the rules of fractions, since we have multiplied both the numerator and the denominator by the same quantity, we have not changed the value of the number.

This gives $\dfrac{5 \times \sqrt{2}}{\sqrt{2} \times \sqrt{2}} = \dfrac{5\sqrt{2}}{2}$ and the denominator is now a rational number.

b) First simplify the denominator and then repeat the process in part **a)**, this time multiplying by $\sqrt{3}$.

$$\frac{7}{\sqrt{12}} = \frac{7}{\sqrt{4 \times 3}}$$

$$= \frac{7}{2\sqrt{3}}$$

$$= \frac{7 \times \sqrt{3}}{2\sqrt{3} \times \sqrt{3}}$$

$$= \frac{7\sqrt{3}}{2 \times 3}$$

$$= \frac{7\sqrt{3}}{6}$$

EXERCISE 3.2

1 Simplify the following, stating whether the result is rational or irrational.

a) $\sqrt{12}$ **b)** $\sqrt{1000}$ **c)** $\sqrt{45}$

d) $\sqrt{300}$ **e)** $\sqrt{75}$ **f)** $\sqrt{8} \times \sqrt{2}$

g) $\sqrt{20} \times \sqrt{18}$ **h)** $\sqrt{20} \div \sqrt{5}$ **i)** $\sqrt{80} \times \sqrt{50}$

j) $\sqrt{75} \times \sqrt{15}$ **k)** $\sqrt{40}$ **l)** $\sqrt{54}$

m) $\sqrt{98}$ **n)** $\sqrt{800}$ **o)** $\sqrt{363}$

p) $\sqrt{27} \times \sqrt{3}$ **q)** $\sqrt{250} \times \sqrt{40}$ **r)** $\sqrt{108} \div \sqrt{12}$

s) $\sqrt{90} \times \sqrt{20}$ **t)** $\dfrac{\sqrt{60} \times \sqrt{20}}{\sqrt{12}}$

2 If $x = 4 + \sqrt{3}$ and $y = 4 - \sqrt{3}$, simplify these.
  a) $x + y$
  b) $x - y$
  c) xy

3 If $x = 5 + \sqrt{7}$ and $y = 3 - \sqrt{7}$, simplify these.
  a) $x + y$
  b) $x - y$
  c) xy

4 If $x = 3 + \sqrt{5}$ and $y = 4 - 3\sqrt{5}$, simplify these.
  a) $x + y$
  b) $x - y$
  c) x^2

5 If $x = 4 + \sqrt{11}$ and $y = 9 - 2\sqrt{11}$, simplify these.
  a) $x + y$
  b) $x - y$
  c) x^2

6 If $x = 5 + 2\sqrt{3}$ and $y = 4 - 3\sqrt{2}$, simplify these.
  a) $x\sqrt{3}$
  b) x^2
  c) y^2

7 If $x = 6 - 2\sqrt{5}$ and $y = 3 - 5\sqrt{3}$, simplify these.
  a) $x\sqrt{5}$ **b)** $y\sqrt{3}$
  c) x^2 **d)** y^2

8 Simplify $\sqrt{2}(5 + 3\sqrt{2})^2$.

9 Show that $(10 - 3\sqrt{7})(10 + 3\sqrt{7})$ is rational, finding its value.

10 Rationalise the denominator in each of these irrational fractions.
  a) $\dfrac{1}{\sqrt{2}}$ **b)** $\dfrac{2}{\sqrt{5}}$
  c) $\dfrac{5}{\sqrt{7}}$ **d)** $\dfrac{11}{\sqrt{18}}$
  e) $\dfrac{9}{\sqrt{20}}$ **f)** $\dfrac{1}{\sqrt{7}}$
  g) $\dfrac{3}{\sqrt{2}}$ **h)** $\dfrac{5}{\sqrt{11}}$
  i) $\dfrac{7}{\sqrt{50}}$ **j)** $\dfrac{9}{\sqrt{32}}$

11 Rationalise the denominator and simplify each of these.
  a) $\dfrac{6}{\sqrt{8}}$ **b)** $\dfrac{6}{\sqrt{300}}$
  c) $\dfrac{12}{\sqrt{75}}$ **d)** $\dfrac{\sqrt{48}}{\sqrt{18}}$
  e) $\dfrac{10}{\sqrt{5}}$ **f)** $\dfrac{15}{\sqrt{50}}$
  g) $\dfrac{20}{\sqrt{32}}$ **h)** $\dfrac{12}{\sqrt{20}}$
  i) $\dfrac{10}{\sqrt{2}}$ **j)** $\dfrac{2}{\sqrt{10}}$
  k) $\dfrac{4}{3\sqrt{10}}$ **l)** $\dfrac{14}{5\sqrt{8}}$
  m) $\dfrac{2\sqrt{3}}{3\sqrt{2}}$ **n)** $\dfrac{12\sqrt{6}}{7\sqrt{15}}$

12 Rationalise the denominator and simplify each of these.
  a) $\dfrac{6 + 3\sqrt{2}}{\sqrt{2}}$ **b)** $\dfrac{15 + \sqrt{5}}{2\sqrt{5}}$
  c) $\dfrac{12 + 3\sqrt{2}}{2\sqrt{3}}$ **d)** $\dfrac{5 + 2\sqrt{3}}{\sqrt{6}}$

EXERCISE 3.2 continued

13 Find an exact expression for the yellow area between these two squares, simplifying as much as possible.

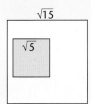
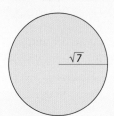

14 Find, as simply as possible, an exact expression for the area of this circle.

15 Find the exact value of x, expressing your answer as simply as possible.

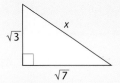
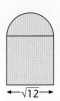

16 Find an exact expression for the total area of this shape formed from a square and a semicircle, simplifying as much as possible.

K KEY IDEAS

- A rational number is a number which can be written as a fraction with both the numerator and the denominator as integers.

- An irrational number is a number which cannot be written as a fraction with both the numerator and the denominator as integers. It is a number which, as a decimal, does not terminate or recur.

- Any recurring decimal can be written as a fraction.

- Surds can be simplified using $\sqrt{a \times b} = \sqrt{a} \times \sqrt{b}$

- Numbers which are the sum of a rational part and irrational part, such as $a + b\sqrt{c}$, can be dealt with using the normal rules of algebra.

- To rationalise a fraction with an irrational denominator, multiply the numerator and the denominator by the surd that is in the denominator.

STAGE

10

4 Trigonometry in non-right-angled triangles

Non-right-angled triangles

All the trigonometry that you have learned so far has been based on finding lengths and angles in right-angled triangles. However, many triangles are not right-angled. Some method needs to be established to find lengths and angles in these other triangles.

It is common practice in this work to use a single letter to represent each side and each angle of the triangle. We use a capital letter to signify an angle and a lower case letter for a side. It is usual for the side opposite an angle to take the same, lower case letter, as shown in the diagram.

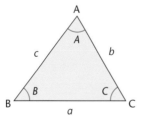

The two major rules for dealing with non-right-angled triangles are called the **sine rule** and the **cosine rule**. It will be obvious why they have these names when you see the formulae.

You do not have to memorise the formulae, as they are on the formulae sheet in the examination, but you do have to know when and how to use them.

The sine rule

A ACTIVITY 1

a) Find c, using right-angled trigonometry. It takes two steps, but is possible!

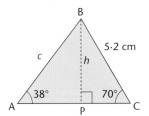

b) Now have a go at a general result, using the same processes as in part **a)**.

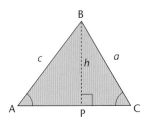

> **EXAM TIP**
> You do not need to be able to prove the sine rule.

(i) Use triangle BCP to find h in terms of a and C.

(ii) Use triangle ABP to find h in terms of c and A.

(iii) Equate your expressions for h and rearrange the result to show that $\dfrac{b}{\sin B} = \dfrac{c}{\sin C}$

You have proved part of the sine rule!

$$\frac{a}{\sin A} = \frac{b}{\sin B} = \frac{c}{\sin C}$$

or $$\frac{\sin A}{a} = \frac{\sin B}{b} = \frac{\sin C}{c}$$

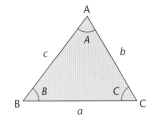

This formula is made up of *three* equal fractions. When using it you take two of the fractions. The two parts are chosen so that there is only one unknown value and three known values. When finding a length the top formula is used; when finding an angle use the bottom one. This will mean that the unknown is on top of the fraction and will make the solution easier.

STAGE
10

4

Use the sine rule

■ when any two angles and one side are known.

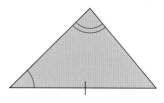

■ when two sides and a non-included angle are known.

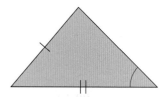

EXAMPLE 1

Find these.

a) Length b

b) Length c

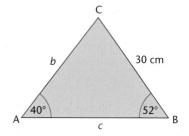

a) Since you are finding a length, choose the formula with lengths on top.
Choose pairs of angles and opposite sides where three of the four values are known and substitute into the formula.

$$\frac{b}{\sin B} = \frac{a}{\sin A}$$

$$\frac{b}{\sin 52°} = \frac{30}{\sin 40°}$$

$$b = \frac{30}{\sin 40°} \times \sin 52°$$

$$b = 36 \cdot 8 \text{ cm to 1 decimal place}$$

b) Before c can be found, angle C is needed.

$$C = 180 - (40 + 52)$$

$$C = 88°$$

$$\frac{c}{\sin C} = \frac{a}{\sin A}$$

$$\frac{c}{\sin 88°} = \frac{30}{\sin 40°}$$

$$c = \frac{30}{\sin 40°} \times \sin 88°$$

$$c = 46 \cdot 6 \text{ cm to 1 decimal place}$$

> **EXAM TIP**
> Though you could use the pair b and B, you should whenever possible use values that are given rather than values that have been calculated.

STAGE
10

▌ EXAMPLE 2

Find the unknown angles and sides in this triangle.

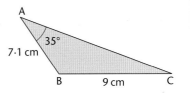

$$\frac{\sin C}{c} = \frac{\sin A}{a}$$

$$\frac{\sin C}{7 \cdot 1} = \frac{\sin 35°}{9}$$

$$\sin C = \frac{\sin 35°}{9} \times 7 \cdot 1$$

$$\sin C = 0 \cdot 4524...$$
$$C = \sin^{-1}(0 \cdot 4524...)$$
$$C = 26 \cdot 9° \text{ to 1 decimal place}$$

The third angle can now be found using the sum of the angles in a triangle.

$$B = 180 - (35 + 26.9)$$

$$B = 118.1° \text{ to 1 decimal place}$$

Notice that it is obtuse (greater than 90°).

You can find the third side using the sine rule again.

$$\frac{b}{\sin B} = \frac{a}{\sin A}$$

$$\frac{b}{\sin 118 \cdot 1°} = \frac{9}{\sin 35°}$$

$$b = \frac{9}{\sin 35°} \times \sin 118 \cdot 1°$$

$$b = 13 \cdot 8 \text{ cm to 1 decimal place}$$

You will learn more about sines and cosines of angles greater than 90° in Chapter 13.

STAGE

10

EXERCISE 4.1

1 Find c, A and a.

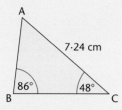

2 Find p, R and r.

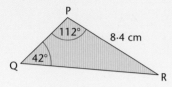

3 Find g, E and e.

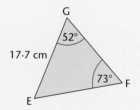

4 Find b, C and c.

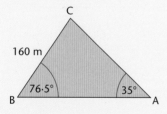

5 Find B, C and c.

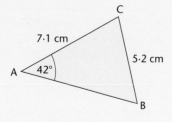

6 Find M, N and n.

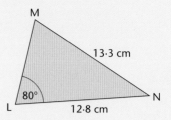

7 Find E, D and d.

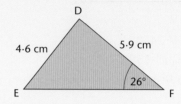

8 Find A, B and b.

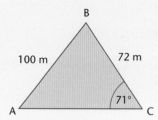

9 Find P, R and r.

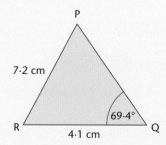

10 Find Y, Z and z.

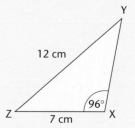

11 Find T, S and s.

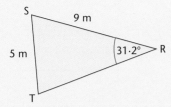

12 Find b, C and c.

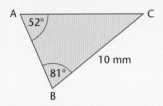

13 Find y, Z and z.

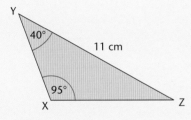

14 Find s, T and t.

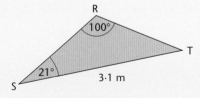

15 In triangle ABC, $c = 23$ cm, $b = 19 \cdot 4$ cm and angle C is $54°$. Find the length of BC.

16 In triangle PQR, $p = 12$ cm, $q = 13 \cdot 4$ cm and angle Q is $56°$. Find the length of PQ.

17 Calculate the largest angle of the triangle ABC given that $A = 35°$, $a = 8 \cdot 9$ cm and $c = 12$ cm.

18 Two points, A and B, are 30 m apart on horizontal ground. The angles of elevation of the top, T, of a vertical tower, TC, from A and B are $27°$ and $40°$ respectively.

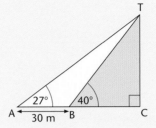

a) Find AT and BT.
b) Find the height of the tower, TC.

19 A river has two parallel banks. Points A and C are on one side of the river, 45 m apart. The angles from these points to a tree, B, on the opposite bank are $68°$ and $34°$, as shown in the diagram.

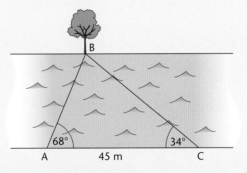

a) Find AB and BC.
b) Find the width of the river.

20 The top of a hill can be seen from A with an angle of elevation $40°$ and from C with an angle of elevation $55°$. AC = 120 m. Calculate the distance AB.

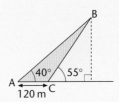

EXERCISE 4.1 continued

21   Town B is 45 km due east of town A. Town C is on a bearing of 057° from town A and a bearing of 341° from town B.

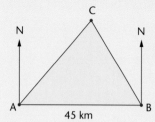

Find how far town C is from towns A and B.

22   A child's slide, RST, is shown in the diagram.

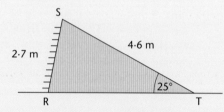

Find the distance RT.

23   ABCDEFGH is a cuboid.
ACH is a triangle contained within the cuboid.
Angle AHC is 80°.

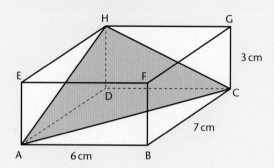

Calculate the size of these angles, giving your answers to the nearest degree.
a)   Angle ACH
b)   Angle CAH

The cosine rule

ACTIVITY 2

a) Another multi-step problem! Use the information in this triangle to work out a. This time you'll need Pythagoras' theorem as well as trigonometry – it is a three-step problem.

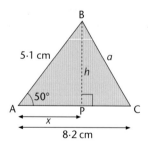

A ACTIVITY 2 continued

b) If you managed the challenge of generalising in part **b)** of Activity 1, try this.

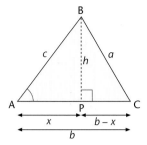

(i) Use triangle BCP to find a^2 in terms of h, b and x.

(ii) Use triangle ABP to find c^2 in terms of h and x.

(iii) Use triangle ABP to find x in terms of c and A.

(iv) Substitute for $(h^2 + x^2)$ and x in your equation for a^2.

You have proved the cosine rule!

$$a^2 = b^2 + c^2 - 2bc\cos A$$
$$b^2 = c^2 + a^2 - 2ca\cos B$$
$$c^2 = a^2 + b^2 - 2ab\cos C$$

or $\cos A = \dfrac{b^2 + c^2 - a^2}{2bc}$

$\cos B = \dfrac{c^2 + a^2 - b^2}{2ca}$

$\cos C = \dfrac{a^2 + b^2 - c^2}{2ab}$

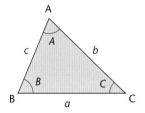

There are three versions of each of the formulae, but notice that they have exactly the same structure and pattern. Once again, there is one form to use when finding length and one for finding angles.

STAGE
10

4

Use the cosine rule

- when all the sides are known.

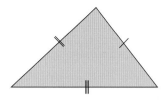

- when two sides and the included angle are known.

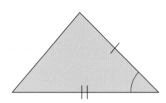

EXAMPLE 3

Find c.

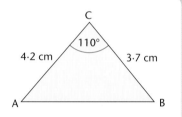

$c^2 = a^2 + b^2 - 2ab\cos C$
$c^2 = 3{\cdot}7^2 + 4{\cdot}2^2 - (2 \times 3{\cdot}7 \times 4{\cdot}2 \times \cos 110°)$
$c^2 = 13{\cdot}69 + 17{\cdot}64 - (^{-}10{\cdot}63...)$
$c^2 = 41{\cdot}96...$
$c = 6{\cdot}48$ cm to 2 decimal places

EXAM TIP

Notice that the cosine of an obtuse angle is negative. Your calculator will give $\cos 110° = {^-}0{\cdot}342....$ More detail is given in Chapter 13.

EXAMPLE 4

Find R.

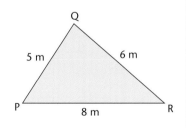

STAGE
10

$\cos R = \dfrac{p^2 + q^2 - r^2}{2pq}$

$\cos R = \dfrac{6^2 + 8^2 - 5^2}{2 \times 6 \times 8}$

$\cos R = 0{\cdot}781\,25...$
$\quad R = \cos^{-1}(0{\cdot}781\,25...)$
$\quad R = 38{\cdot}6°$ to 1 decimal place

EXERCISE 4.2

1 Find *a*.

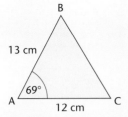

2 Find *c*.

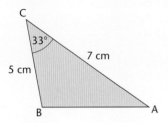

3 Find *p*.

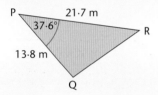

4 Find *r*.

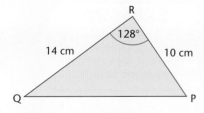

5 Find *C*.

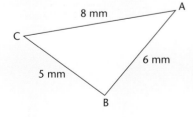

6 Find *B*.

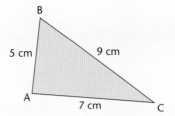

7 Find *G*.

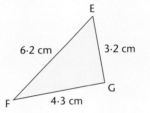

8 Find length BC.

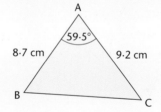

9 Find *a*.

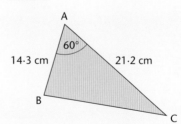

10 Find length PQ.

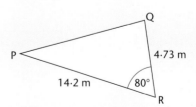

11 Find *s*.

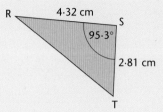

12 Find angle ABC.

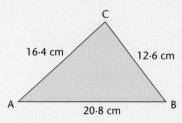

13 Find angle *B*.

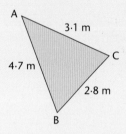

14 Three towns, A, B and C, are shown in the diagram.

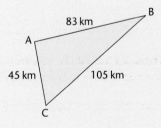

Find the angle ACB.

15 Find the smallest angle in this triangle.

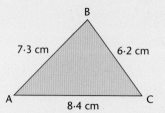

16 Three towns, A, B and C, are positioned as shown in the diagram. Find the three angles inside the triangle formed.

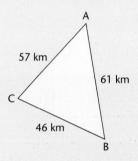

17 A cross-country runner runs 4 km due north and then 6·7 km in a south-east direction.
How far is she from her starting point?

> **EXAM TIP**
>
> Always draw a sketch if you are not given a diagram. Label the sides and angles with the information you have.

18 From a boat C, A is 9 km away on a bearing of 058° and B is 12 km away on a bearing of 110°.
Calculate the distance AB.

19 The diagram shows the cross-section of a roof.
The house is 12 m wide.

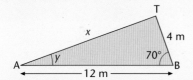

Calculate the length of the roof, x, and the angle of the slope, y.

20 A parallelogram has sides of length 5·1 cm and 2·5 cm.
Adjacent sides are separated by an angle of 70°.
Find the length of each of the diagonals of the parallelogram.

21 A vertical flagpole, FP, 25 m high, stands on horizontal ground.
It is supported by two ropes, FA and FB. FA = 35 m and FB = 40 m.
The angle, AFB, between the ropes is 75°.

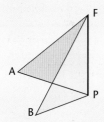

a) Work out the length AB.
b) Find angle APB.

22 Three points, A, B and C, form an equilateral triangle on horizontal ground with sides 10 m.
Vertical posts PA, QB and RC are placed in the ground.
PA = 4 m, QB = 10 m and RC = 6 m.

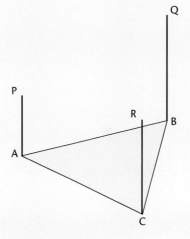

a) **(i)** Find PQ.
 (ii) Find PR.
 (iii) Find QR.
b) Hence find the size of
 (i) angle RPQ.
 (ii) angle PRQ.

STAGE
10

203

Trigonometry in non-right-angled triangles

The general formula for the area of any triangle

A | ACTIVITY 3

a) Just a two-step problem this time.
Find the area of this triangle, ABC.

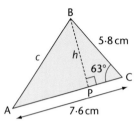

b) A much easier generalisation this time than in Activities 1 and 2.
 (i) Use triangle BCP to find h in terms of a and C.
 (ii) Use this expression for h to show that the area of triangle ABC is $\frac{1}{2}ab\sin C$.

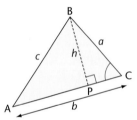

Once again, there are three versions of the formula, and again the letters have a 'circular' structure. Each formula requires two adjacent sides and the included angle.

Area of triangle ABC $= \frac{1}{2}\,ab\,\sin C$

$\qquad\qquad\qquad\quad = \frac{1}{2}\,bc\,\sin A$

$\qquad\qquad\qquad\quad = \frac{1}{2}\,ca\,\sin B$

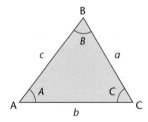

STAGE
10

204

EXAMPLE 5

Find the area of the triangle shown.

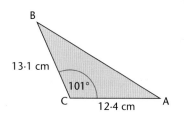

Area $= \frac{1}{2}ab\sin C$

$\qquad = \frac{1}{2} \times 13\cdot1 \times 12\cdot4 \times \sin 101°$

$\qquad = 79\cdot7 \text{ cm}^2$ to 1 decimal place

EXERCISE 4.3

1 Find the area of each of these triangles.

a)

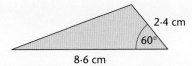

b)

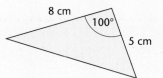

c)

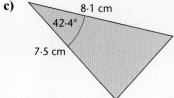

d)

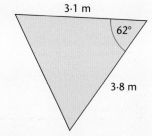

e)

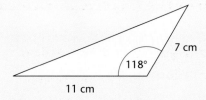

f)

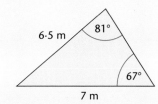

STAGE

10

2 In triangle ABC,
$a = 10\,\text{cm}$, $c = 6\,\text{cm}$ and $B = 150°$.
Find the area of the triangle.

3 In triangle ABC, $a = 4\,\text{cm}$, $c = 7\,\text{cm}$ and
its area is $13\cdot4\,\text{cm}^2$.
Find the size of angle B.

4 The area of triangle PQR is $273\,\text{cm}^2$.
Given that PQ = $12\cdot8\,\text{cm}$ and angle
PQR = $107°$, find QR.

5 Calculate the area of parallelogram
ABCD in which AB = $6\,\text{cm}$, BC = $9\,\text{cm}$
and angle ABC = $41\cdot4°$.

EXAM TIP

Remember to draw a
diagram and label it
with the information
you know.

KEY IDEAS

■ Use the sine rule to calculate lengths and angles in non-right-angled triangles

■ when any two angles and one side are known.

■ when two sides and a non-included angle are known.

■ Use the cosine rule to calculate lengths and angles in non-right-angled triangles

■ when all three sides are known.

■ when two sides and the included angle are known.

■ To calculate the area of a triangle when you don't know the height, use

Area = $\frac{1}{2}ab\sin C$

or its equivalent.

STAGE
10

Revision exercise A1

1 a) Draw the graphs of $y = x^2 - 2x + 3$ and $y = 4x + 1$ on the same grid. Use values of x from 0 to 6.

b) Use the graphs to solve, simultaneously, the equations $y = x^2 - 2x + 3$ and $y = 4x + 1$. Give the answers correct to 1 decimal place.

2 a) Draw the graphs of $x^2 + y^2 = 9$ and $y = x + 2$ on the same grid. Use a scale of 1 cm to 1 unit for both x and y.

b) Use the graph to solve the simultaneous equations $x^2 + y^2 = 9$ and $y = x + 2$. Give the answers correct to 1 decimal place.

3 Solve these simultaneous equations graphically.
$x^2 + y^2 = 36$ and $y = x + 6$

4 a) Draw the graph of $y = x^2 + 3x - 7$ for x from $^-6$ to 3.

b) Use the graph to solve these equations.
(i) $x^2 + 3x - 7 = 0$
(ii) $x^2 + 3x - 10 = 0$

c) (i) Find the line that must be drawn to solve $x^2 + x - 4 = 0$.
(ii) Draw the line and use it to solve $x^2 + x - 4 = 0$.

5 In a chemical reaction, the mass of a chemical present is decreasing by 5% per minute.
Initially there is 20 g of the chemical.
Find, in minutes correct to 1 decimal place, the time that passes before there is 2 g of this chemical left.

6 Copy and complete the table of values for $y = 2^{-x}$.

x	0	0·5	1	1·5	2	2·5	3	3·5	4
y	1								

Plot the graph of $y = 2^{-x}$ for these values.
Use a scale of 2 cm to 1 unit on the x-axis and 1 cm to 0·1 unit on the y-axis.
Use your graph to estimate
a) the value of y when $x = 1·8$.
b) the solution to the equation $2^{-x} = 0·6$.

7 The size, y, of a population of flies after t days was given by $y = 100 \times 1·2^t$.
a) What was the size of the population at $t = 0$?
b) What was the size of the population after 5 days?
c) Use trial and improvement to find the number of days it took for the population to reach 1000, assuming this rate of growth continued. Give your answer to 1 decimal place.

8 The curve $y = ab^x$ passes through $(0, 10)$ and $(2, 6·4)$.
Find the values of a and b.

9 State which of these numbers are rational and which are irrational, showing how you decide.
a) $^-1·6$
b) $0·\dot{7}\dot{3}$
c) $\dfrac{5\pi}{3}$
d) $7 + 2\sqrt{3}$
e) $1·414$

10 Convert these fractions to recurring decimals, using the dot notation.
a) $\dfrac{5}{11}$
b) $\dfrac{212}{999}$
c) $\dfrac{37}{495}$

STAGE
10

11 Convert these recurring decimals to fractions or mixed numbers in their lowest terms.

a) $0 \cdot \dot{5} \dot{4}$ b) $3 \cdot 1 \dot{4} \dot{7}$

c) $0 \cdot \dot{2} 03 \dot{4}$

Do not use a calculator for questions **12** to **16**.

12 Simplify these.

a) $\sqrt{32}$

b) $\sqrt{150}$

c) $\sqrt{128}$

d) $\sqrt{12} \times \sqrt{75}$

e) $\sqrt{10} \times \sqrt{18}$

f) $\sqrt{72} \div 3$

g) $\sqrt{288} \times \sqrt{48}$

13 If $x = 3 + \sqrt{7}$ and $y = 5 - 4\sqrt{7}$, simplify these.

a) $x + y$ b) $x - y$

c) xy

14 If $x = 5 + 2\sqrt{3}$ and $y = 5 - 2\sqrt{3}$, simplify these.

a) x^2 b) y^2

c) xy

15 Simplify $\sqrt{10}\left(5 + 2\sqrt{10}\right)^2$.

16 Rationalise the denominator in the following, simplifying where possible.

a) $\dfrac{11}{\sqrt{2}}$ b) $\dfrac{15}{\sqrt{12}}$ c) $\dfrac{6}{\sqrt{27}}$

17 In triangle PRT, find

a) angle PRT.

b) angle PTR.

c) PR.

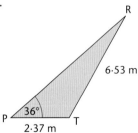

18 In triangle ABC, find

a) BC.

b) angle ABC.

c) the area of triangle ABC.

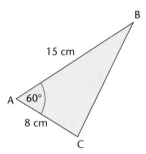

19 In triangle KLM, find

a) angle LKM.

b) angle KML.

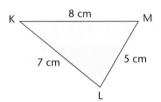

20 In triangle ABC, find

a) AC.

b) angle BAC.

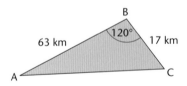

21 ABCD is a field with dimensions as shown in the diagram. Calculate the area of the field.

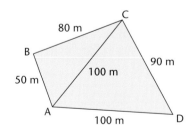

Trends and time series

You will learn about

- Identifying seasonality and trends in time series
- Interpreting graphs that model real situations

You should already know

- How to plot time series graphs
- How to calculate moving averages
- How to interpret simple graphs of everyday situations, such as a journey

This chapter consolidates and extends the work you have done on interpreting time series and graphs that model real-life situations.

EXAMPLE 1

Heritage Holidays specialise in holidays to Egypt, Italy and Greece.

The table shows the number of people taking their holidays with the company.

Year	Quarter	Number of people (100s)	Four-point moving average
2001	Spring	4	
	Summer	8	
			4·5
	Autumn	3	
			5·25
	Winter	3	
			6·5
2002	Spring	7	
			7·25
	Summer	13	
			8
	Autumn	6	
			8·5
	Winter	6	
			9
2003	Spring	9	
			9·5
	Summer	15	
			9·75
	Autumn	8	
			10
	Winter	7	
			10·75
2004	Spring	10	
			11·75
	Summer	18	
			12·5
	Autumn	12	
			13·5
	Winter	10	
			14
2005	Spring	14	
			14·75
	Summer	20	
			14·75
	Autumn	15	
	Winter	10	

▌▌EXAMPLE 1 continued

These numbers have been plotted on a graph together with the four-point moving averages.

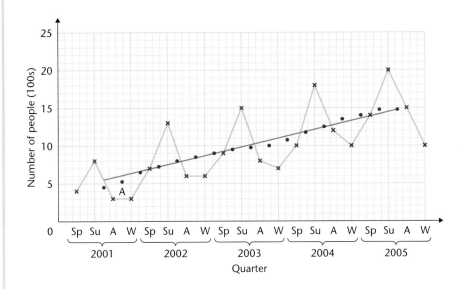

a) One of the moving average points has been marked A.
Show how this has been calculated.

b) A line of best fit, a trendline, has been drawn through the moving average points.
Use this line to describe the trend.

c) Use the trendline to estimate the number of bookings for the spring of 2006.

a) The point A is the moving average for last three quarters of 2001 and the first quarter of 2002.

Its value is $\dfrac{8 + 3 + 3 + 7}{4} = 5.25$

b) The trend is a steady increase each year.

c) Extending the trendline to the next moving average point, midway between autumn and winter 2005, gives a value of 15·5.

This value would be obtained by adding the values for summer, autumn and winter 2005 and spring 2006, and dividing by 4. If x is the value for spring 2006 then

$$\frac{20 + 15 + 10 + x}{4} = 15.5$$

$$45 + x = 62$$

$$x = 17$$

An estimate of the number of bookings for the spring of 2006 is 1700.

STAGE
10

Informal methods like that used in Example 1, part **c)**, are often used to estimate future performance, but of course this may be affected by unknown factors. Do not rely too much on forecasts!

A ACTIVITY 1

Look on the internet to find other graphs showing time series and trends or seasonal variation.

Good sites to look at are those of government statistics showing social trends, or those with weather records.

EXAMPLE 2

This graph shows the velocity of a car.

a) Calculate the gradient of this graph and say what it represents.

b) Sketch a graph to show the shape of the graph of the distance travelled by the car plotted against the time.

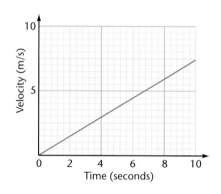

a) Gradient = $\dfrac{7.5 \, \text{m/s}}{10 \, \text{s}}$ = $0.75 \, \text{m/s}^2$

This represents the acceleration of the car.

b) As the velocity increases, the car goes further in any given period of time.

The graph looks like this.

> ### EXAM TIP
> Always include the units when working out the gradients of graphs showing practical situations. The units help you to see what the gradient represents.

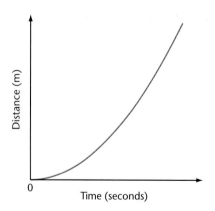

EXERCISE 5.1

1 The table shows the daily sales figures in a clothing shop over a 4-week period.

	M	Tu	W	Th	F	Sa	Su
Week 1	1256	1785	875	2564	1932	3954	2703
Week 2	1307	2013	1076	2197	2102	4023	2893
Week 3	1293	1847	1132	1988	2234	4456	3215
Week 4	1402	1951	1207	2394	1987	5201	2694

a) Plot these figures on a graph.
b) Calculate the 7-day moving averages.
c) Plot the moving averages and add a trendline if appropriate.
d) Comment on any trend and daily variation.

2 a) Calculate the last moving average for this graph, which has not been plotted.

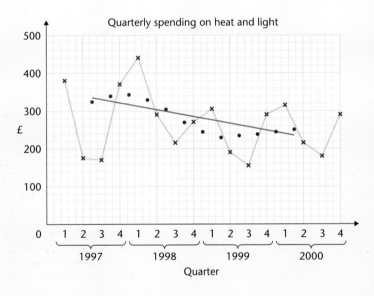

b) Comment on the seasonal variation and trends in the spending shown on this graph.

3 Paul checks his gas bill each quarter. The quarterly amounts are shown in the table.

Year	Quarter	Amount (£, to the nearest £)
2003	1	174
	2	55
	3	62
	4	185
2004	1	193
	2	56
	3	63
	4	225
2005	1	199
	2	83
	3	95
	4	192
2006	1	230

a) Plot these figures on a graph.
b) Calculate the 4-point moving averages.
c) Plot the moving averages and add a trendline.
d) Comment on the variation in the actual values and on the trendline.
e) Use your trendline to estimate the bill for the 2nd quarter in 2006.

4 The table shows the daily attendance figures at a cinema over a 4-week period.

	M	Tu	W	Th	F	Sa	Su
Week 1	137	201	414	216	521	823	411
Week 2	203	156	304	242	487	615	273
Week 3	166	231	322	284	534	725	389
Week 4	189	177	385	302	498	782	450

a) Plot these figures on a graph.
b) Calculate the 7-day moving averages.
c) Plot the moving averages and add a trendline if appropriate.
d) Comment on any trend and daily variation.

5 An insurance company gave these bonuses on the sums assured.

	1997	1998	1999	2000	2001	2002	2003	2004	2005	2006
Bonus (%)	4·40	3·84	3·36	3·24	3·24	3·24	3·24	2·82	2·40	1·92

a) Comment on the trends.

b) Giving your reasons, forecast the bonus for 2007.

6 The table below shows the average temperature (in °C) for Westfield-on-Sea over a 5-day period.

	12–4 a.m.	4–8 a.m.	8–12 p.m.	12–4 p.m.	4–8 p.m.	8–12 p.m.
Monday	⁻2	2	6	8	4	0
Tuesday	⁻1	2	8	7	5	3
Wednesday	2	3	7	9	5	2
Thursday	4	6	9	11	6	5
Friday	5	6	8	9	5	4

Calculate the 6-point moving averages and plot the graph.
Comment on the weather pattern seen.

7 The graph shows a company's quarterly spending on photocopying.

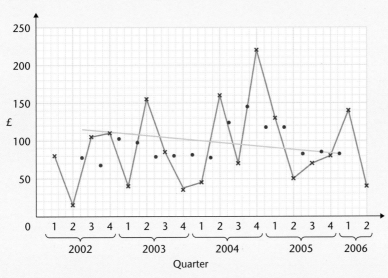

a) Show how the highest point on the moving averages has been calculated. Explain what factors on the graph of monthly spending have caused this to be the highest point.

b) Describe the trend.

c) Estimate the spending on photocopying for the next quarter.

d) Do you think your estimate in part **c)** is a good one? Explain your answer.

STAGE

10

8 This graph shows the speed of a ball bearing falling through a cylinder filled with thick oil.

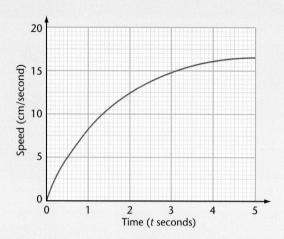

Describe what this graph shows about the acceleration of the ball bearing.

9 These containers are filled at a constant rate.

a)    b)    c) 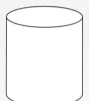   d)

Sketch a graph for each one, showing how the depth in the container varies with time.

10 The graph models water pouring into a bath.
   a) Describe what is happening at points A, B, C, D and E.
   b) Find the rate of flow at A.

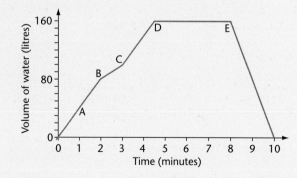

EXERCISE 5.1 continued

11 This graph shows a car moving with constant velocity.
Draw the graph of the distance travelled plotted against time.

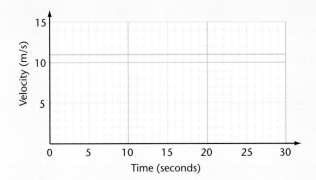

12 The graph shows a lorry travelling uphill on a motorway.
Calculate the deceleration of the lorry.

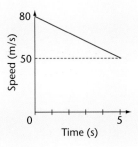

K KEY IDEAS

- A time series shows the variation of a set of figures over periods of time. These periods can be quarterly, daily, monthly, etc. They are usually displayed on a graph.

- Moving averages can be used to help identify trends in the data.

- To calculate a moving average, for example for quarterly figures, first calculate the mean for the first four quarters. Then omit the first quarter and include the fifth quarter, and find the new mean. Then omit the second quarter and include the sixth, and so on.

- A moving average is plotted in the middle of the interval to which it relates.

- The gradient of a distance–time graph gives the velocity.

- The gradient of a velocity–time graph gives the acceleration.

6 Congruency – proving and using

You will learn about

- Proving that triangles are congruent
- Verifying the standard ruler and compass methods of construction
- Using congruency to show that translation, rotation and reflection preserve length and angle size

You should already know

- How to draw the perpendicular bisector of a line
- How to draw an angle bisector

 ACTIVITY 1

Work in pairs.

a) Each draw a triangle with sides of 3 cm, 4 cm and 5 cm.

Are your triangles congruent?

b) Each draw a triangle, ABC, with AB = 6 cm, A = 60° and BC = 5·8 cm.

Are your triangles congruent?

If they are, can you draw one which isn't?

c) Try to find other examples of information which can lead to:
   (i) one triangle only.
   (ii) two possible triangles.

STAGE
10

For any two triangles to be congruent, they must be the same shape and the same size. This means that they will fit exactly on to each other when one of them is rotated, reflected or translated.

> **Thus, it can be seen that if two triangles are congruent then**
> - **the three angles in one triangle equal the corresponding three angles of the other triangle (also true for similar triangles).**
> - **the three sides in one triangle equal the corresponding three sides of the other triangle (not true for similar triangles).**

Two triangles are congruent if any of the four sets of conditions below is satisfied.

Two sides and the included angle of one triangle are equal to two sides and the included angle of the other triangle (side, angle, side or SAS).

The three sides of one triangle are equal to the three sides of the other triangle (side, side, side or SSS).

Two angles and a side in one triangle are equal to the corresponding two angles and side in the other triangle (angle, side, angle or ASA).

Both triangles are right-angled and the hypotenuse and one side of one triangle are equal to the hypotenuse and corresponding side of the other triangle (right angle, hypotenuse, side or RHS).

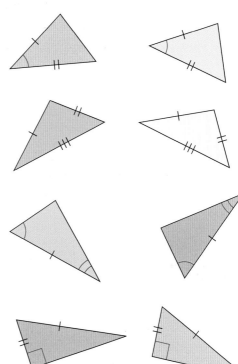

Note that the test involving two sides and an angle is valid only if the angle is the included angle. If the given angle is not the included angle then there are two possible solutions – the ambiguous case which you may have met when studying the sine rule, or in Activity 1 in this chapter.

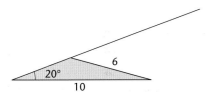

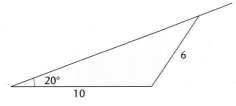

Note that the test involving two angles and a side is valid for any pair of corresponding sides. Once you know two angles of a triangle you can find the third. Using the side between two angles helps ensure that you are comparing corresponding sides.

EXAMPLE 1

Prove that the diagonal of a parallelogram splits the parallelogram into two congruent triangles.

Hence show that the opposite sides of a parallelogram are equal.

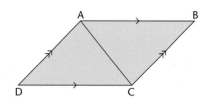

Angle BAC = angle ACD   (Alternate angles, AB and DC parallel.)
Angle ACB = angle DAC   (Alternate angles, AD and BC parallel.)

In triangles ACB and ACD, AC is common.

Hence these triangles are congruent (ASA).

Since the triangles are congruent, corresponding pairs of sides are equal.

So CB = AD
and AB = CD.

That is, the opposite sides of a parallelogram are equal.

EXAMPLE 2

Use congruency to show that the reflection A'B' of a line AB in any line has a length equal to AB.

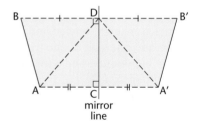
mirror line

Let AA' and BB' cross the mirror line at C and D respectively.

By definition of reflection, AC = A'C, and AA', crosses the mirror line at right angles.

So considering triangles ACD and A'CD, since CD is common, these triangles are congruent (SAS).

This implies AD = A'D and angle ADC = angle A'DC.

Considering triangles BDA and B'DA',

by definition of reflection BD = B'D and angle BDC = angle B'DC = 90°,

so angle BDA = angle B'DA' = 90° – angle CDA.

So triangles BDA and B'DA' are congruent and, in particular, AB = A'B' as required.

Example 2 is a logical proof. It builds up from basic definitions. You might like to try adapting the proof for when the mirror line crosses AB or goes through A or B.

EXERCISE 6.1

1 Which of these triangles are congruent?
Give a reason or explanation for your answer.

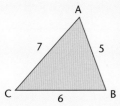 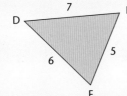 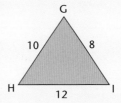 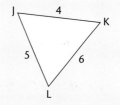

2 Which of these pairs of triangles are congruent?
Explain your answer in each case.

a)

b)

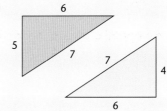

c)

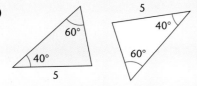

d)

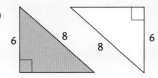

e)

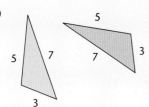

f)

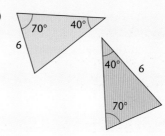

STAGE
10

221

3 Which of the triangles **(i) (ii) (iii) (iv) (v) (vi)** are congruent to any of triangles **A**, **B** or **C**? Explain your answer in each case.

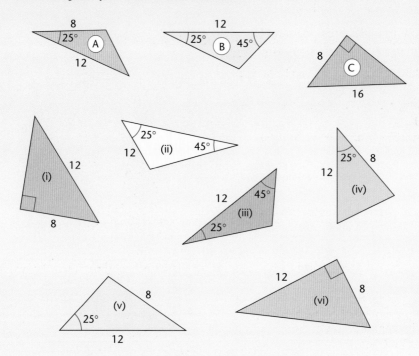

4 Prove that diagonal AC bisects the angles at A and C.

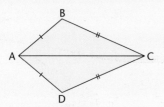

A sketch will prove useful for question **5** to **11**.

5 Prove that the angles opposite the equal sides in an isosceles triangle are equal.

6 Sketch an isosceles triangle ABC with the equal angles at B and C.
Draw a straight line from the midpoint of side BC to the vertex A.
Prove that this line bisects angle BAC and is perpendicular to side BC.

7 Prove that the diagonals of a rectangle are equal in length.

8 Use the properties of congruent triangles to prove that the diagonals of a rhombus bisect each other at right angles.

9 Prove that the diagonals of a kite bisect the shorter diagonal at right angles and bisect the unequal pair of angles.

10 Sketch an equilateral triangle.
Join the midpoints of each side to make a second triangle.
Prove that this triangle is equilateral.

11 Prove that the centre, O, of a rotation which maps AB on to A′B′ is where the perpendicular bisectors of AA′ and BB′ meet.
Prove also that AB is equal in length to A′B′.

12 In the diagram, AB = AC, angle XAE = 140°
and CE = CF.
Find angle ADE, giving reasons for your
answer.

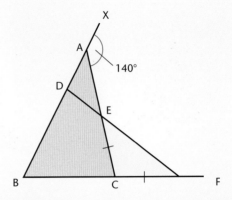

13 In this diagram, CD is the bisector of angle
ACB and AE is parallel to DC.
Prove that AC = CE.

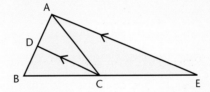

14 In this diagram, AB = AC, CD is parallel to
BA, CD = CB and angle BAC = 40°.
Calculate the size of angle DBC, giving
reasons for your answer.

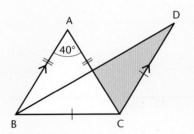

15 In this diagram, angle CBX = 122°, angle
DEC = 33° and AX is parallel to DY.
Calculate the size of angles BCY and BAD,
giving reasons for your answers.

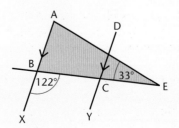

EXERCISE 6.1 continued

16 In triangle ABC, BD bisects angle ABC,
angle ACB = 80° and BD = BC.
Prove that angle BAC = 60°.

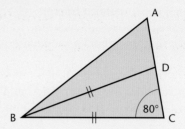

17 ABC is a right-angled triangle.
Angle ACB = 55° and angle PQC = 90°.
Prove that angle a = 145°.

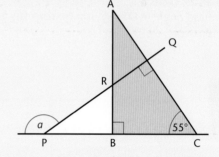

ACTIVITY 2

a) Use ruler and compasses to draw the perpendicular bisector of a line AB 10 cm long.

b) Mark any point on the bisector P.

Join P to A and B.

c) Use congruent triangles to show why the construction works.

ACTIVITY 3

a) Draw an angle of 110° and lable it ABC.
Use ruler and compasses to draw the bisector of angle ABC.

b) Use congruent triangles to show why the construction works.

K KEY IDEAS

Two triangles are congruent if any of these four sets of conditions is satisfied.

■ The corresponding sides of each triangle are equal (SSS).

■ Two sides and the included angle in each triangle are equal (SAS).

■ Two angles and the corresponding side in each triangle are equal (ASA).

■ Both triangles are right-angled, the sides opposite the right angles are equal (i.e. the hypotenuse in each triangle) and another pair of sides are equal (RHS).

7 Calculating the roots of equations

You will learn about

- Solving quadratic equations by completing the square
- Solving quadratic equations using the formula

You should already know

- How to expand and simplify $(ax + b)(cx + d)$
- How to factorise quadratic expressions
- How to solve quadratic equations using factorising or graphs

Completing the square

This quadratic equation factorises.

$$x^2 - 4x + 3 = 0$$

This one does not.

$$x^2 - 4x + 1 = 0$$

In Chapter 1 you learned how to use graphs to solve equations like this, by drawing the graph of $y = x^2 - 4x + 1$ and reading off the values of x where $y = 0$.

In this chapter we look at two methods of calculating the **roots** of the equation. The roots of an equation are the values for which it is true.

A ACTIVITY 1

Expand and simplify these.

$(x + 1)^2 = (x + 1)(x + 1) =$

$(x - 2)^2 = (x - 2)(x - 2) =$

$(x + 3)^2 =$

$(x - 5)^2 =$

$(2x + 1)^2 =$

$(3x - 2)^2 =$

$(5x + 4)^2 =$

Spot the patterns in your answers and just write down the answers to these.

$(x - 4)^2$

$(4x + 1)^2$

$(2x - 3)^2$

A ACTIVITY 2

Copy and complete these factorisations.

$x^2 + 4x + \ldots = (x + 2)^2$

$x^2 + \ldots + 9 = (x + 3)^2$

$x^2 + 10x + 25 = (x + \ldots\ )^2$

$x^2 + 2x + \ldots = (x + \ldots)^2$

$x^2 + 12x + \ldots = (x + \ldots)^2$

$x^2 - 6x + \ldots = (x - \ldots)^2$

$x^2 - 16x + \ldots = (x \ldots\ldots)^2$

$x^2 - 7x + \ldots = (x \ldots\ldots)^2$

$x^2 + 5x + \ldots = (x + \ldots)^2$

$4x^2 + 12x + \ldots = (2x + \ldots)^2$

$9x^2 + 6x + \ldots = (3x + \ldots)^2$

$6x^2 + 24x + \ldots = 6(x + \ldots)^2$

STAGE
10

$x^2 - 4x$ is part of the expansion $(x - 2)^2 = x^2 - 4x + 4$.

In this method for solving a quadratic equation we 'complete the square' so that the left-hand side of the equation is a square $(mx + k)^2$.

EXAMPLE 1

Solve $x^2 - 4x + 1 = 0$.

$$x^2 - 4x + 1 = 0$$

$$x^2 - 4x + 1 + 3 = 0 + 3$$

Add a number so the left-hand side is a complete square.

$$x^2 - 4x + 4 = 3$$

$$(x - 2)^2 = 3$$

Factorise the left-hand side.

$$x - 2 = \sqrt{3} \text{ or } {}^{-}\sqrt{3}$$

Take the square root of both sides.

$$x = 2 + \sqrt{3} \text{ or } 2 - \sqrt{3}$$

This is usually written $x = 2 \pm \sqrt{3}$

$x = 3{\cdot}73$ or $0{\cdot}27$ to 2 decimal places

EXAMPLE 2

Solve $x^2 + 3x - 5 = 0$.

$$x^2 + 3x - 5 = 0$$

$$x^2 + 3x + \frac{9}{4} = 5 + \frac{9}{4}$$

$$\left(x + \frac{3}{2}\right)^2 = \frac{29}{4}$$

$$x + \frac{3}{2} = \pm\sqrt{\frac{29}{4}}$$

$$x = \frac{{}^{-}3}{2} \pm \sqrt{\frac{29}{4}}$$

$x = 1{\cdot}19$ or ${}^{-}4{\cdot}19$ to 2 decimal places

An alternative method to avoid fractions is to multiply the equation through by 4 at the start.

Hint:
Here is a method to work out how much to add.
Since $(mx + k)^2 = m^2x^2 + 2kmx + k^2$, use the coefficient of x^2 to see that $m = 2$ then use the coefficient of x to see that $2km = 12$, so $k = 3$.
We need $k^2 = 9$ on the LHS.

EXAMPLE 2 alternative method

Solve $x^2 + 3x - 5 = 0$.

$x^2 + 3x - 5 = 0$

$4x^2 + 12x - 20 = 0$ Multiply by 4.

$4x^2 + 12x + 9 = 29$ Add 29 so the LHS is a complete square.

$(2x + 3)^2 = 29$

$2x + 3 = \pm\sqrt{29}$ Take the square root.

$2x = ^-3 \pm\sqrt{29}$ Rearrange to make x the subject.

$x = \dfrac{^-3 \pm\sqrt{29}}{2}$

$x = 1 \cdot 19$ or $^-4 \cdot 19$ to 2 decimal places

If the coefficient of x^2 is not a square already, multiply the equation through at the start to make it so.

EXAMPLE 3

Solve $2x^2 + 10x + 5 = 0$.

$2x^2 + 10x + 5 = 0$

$4x^2 + 20x + 10 = 0$ Multiply by 2.

$4x^2 + 20x + 25 = 15$ Add 15 to get 25 on the LHS.

$(2x + 5)^2 = 15$

$2x + 5 = \pm\sqrt{15}$

$2x = ^-5 \pm\sqrt{15}$

$x = \dfrac{^-5 \pm\sqrt{15}}{2}$

$x = ^-0 \cdot 56$ or $^-4 \cdot 44$ to 2 decimal places

A ACTIVITY 3

a) What is the smallest possible value of $(x - 3)^2$?
What would be the value of x in that case?

b) What is the smallest possible value of $(x - 3)^2 + 5$?

c) What is the smallest possible value of $(x - 3)^2 - 2$?

d) What is the smallest possible value of $(x + 2)^2 + 4$?
What would be the value of x in that case?

e) What is the smallest possible value of $(x - 1)^2 - 6$?
What would be the value of x in that case?

f) What is the smallest possible value of $(x + m)^2 + n$?
What would be the value of x in that case?

EXERCISE 7.1

Solve the equations in questions **1** to **20** by completing the square.

1 $x^2 - 6x + 4 = 0$

2 $x^2 - 8x - 2 = 0$

3 $x^2 + 10x + 5 = 0$

4 $x^2 + 3x - 1 = 0$

5 $x^2 - 7x + 2 = 0$

6 $x^2 - 2x - 2 = 0$

7 $x^2 + 6x - 4 = 0$

8 $x^2 + 8x + 3 = 0$

9 $x^2 - 10x + 6 = 0$

10 $x^2 + 3x - 5 = 0$

11 $x^2 - 5x + 2 = 0$

12 $4x^2 + 3x - 2 = 0$

13 $4x^2 - 6x + 1 = 0$

14 $3x^2 - 4x - 5 = 0$

15 $2x^2 + 12x + 3 = 0$

16 $3x^2 + 2x - 2 = 0$

17 $5x^2 - 6x - 4 = 0$

18 $2x^2 - 5x - 2 = 0$

19 $5x^2 - x - 1 = 0$

20 $16x^2 + 12x + 1 = 0$

21 **a)** Write $x^2 + 12x + 12$ in the form $(x + m)^2 - n$.
   b) Hence state the minimum value of $y = x^2 + 12x + 12$.
   c) Using your answer to part **a)**, or otherwise, solve the equation $x^2 + 12x + 12 = 0$.
   Leave your answer in the form $p \pm q\sqrt{6}$.

Using the quadratic formula

Using the method of completing the square it may be shown that

The equation $ax^2 + bx + c = 0$ has roots $x = \dfrac{-b \pm \sqrt{b^2 - 4ac}}{2a}$.

CHALLENGE 1

Try proving this formula for yourself.

Hint:
Multiply through by $4a$ to avoid fractions.

If you are not specifically asked to use another method, just to solve the equation, you may quote and use this formula. You are not expected to remember it, you will be given the formula in an examination.

EXAMPLE 4

Solve the equation $x^2 - 4x + 2 = 0$.

Give your answer to 2 decimal places.

In the equation $x^2 - 4x + 2 = 0$, $a = 1$, $b = {}^-4$, $c = 2$.

$$x = \frac{-b \pm \sqrt{b^2 - 4ac}}{2a}$$

$$x = \frac{{}^+4 \pm \sqrt{16 - 4 \times 1 \times 2}}{2}$$

$$= \frac{{}^+4 \pm \sqrt{8}}{2}$$

$$= \frac{4 \pm 2{\cdot}828}{2}$$

$x = 3{\cdot}41$ or $0{\cdot}59$ to 2 decimal places.

> **EXAM TIP**
> Make sure you always include the correct sign when substituting for a, b and c.

> **EXAM TIP**
> It is sensible to do all the working out on a calculator, but it is vital to write down the expression you are going to work out.

STAGE
10

EXAMPLE 5

Solve the equation $3x^2 + 4x - 2 = 0$.

Give your answers to 2 decimal places.

In the equation $3x^2 + 4x - 2 = 0$, $a = 3$, $b = 4$, $c = {}^-2$.

$$x = \frac{{}^-b \pm \sqrt{b^2 - 4ac}}{2a}$$

$$x = \frac{{}^-4 \pm \sqrt{16 - 4 \times 3 \times ({}^-2)}}{6}$$

$$= \frac{{}^-4 \pm \sqrt{16 + 24}}{6}$$

$$= \frac{{}^-4 \pm \sqrt{40}}{6}$$

$$= \frac{{}^-4 \pm 6 \cdot 324}{6}$$

$x = 0 \cdot 39$ or ${}^-1 \cdot 72$ to 2 decimal places

EXAM TIP

The main errors that occur in using the formula are
- errors with the signs, especially with ${}^-4ac$.
- failure to divide the whole expression by $2a$.

EXERCISE 7.2

Use the formula to solve the equations in questions **1** to **14** .
Give your answers correct to 2 decimal places.

1 $x^2 + 8x + 6 = 0$

8 $x^2 + 7x + 4 = 0$

2 $2x^2 - 2x - 3 = 0$

9 $2x^2 - 3x - 4 = 0$

3 $3x^2 + 5x - 1 = 0$

10 $3x^2 + 2x - 2 = 0$

4 $5x^2 - 12x + 5 = 0$

11 $5x^2 - 13x + 7 = 0$

5 $5x^2 + 9x - 6 = 0$

12 $5x^2 + 9x + 3 = 0$

6 $x^2 - 5x - 1 = 0$

13 $7x^2 - 5x - 1 = 0$

7 $3x^2 + 9x + 5 = 0$

14 $3x^2 + 2x - 7 = 0$

EXERCISE 7.2 continued

15–20 Solve the equations in Exercise 7.1 questions **15–20** but this time using the formula.

21 A garden is 8 m longer than it is wide and it has an area of $25\,\text{m}^2$.

Write down an equation and solve it to find the dimensions correct to the nearest centimetre.

22 Dad is three times as old as Tom. Tom works out that in five years' time the product of their ages will be roughly 1000.
Write down an equation and solve it to find out how old they are now.

23 A pen is constructed along an existing wall using 20 m of fencing.

a) If the width is x m, write down an expression for the area enclosed.

b) Write down an equation and solve it to find the dimensions to give an area of $40\,\text{m}^2$.

c) By completing the square, find the maximum area possible with this amount of fencing.

24 A rectangular lawn measuring 22 m by 15 m is surrounding by a path x m wide. Form a simplified expression for the total area of lawn and path.
Write down an equation and solve it to find the width of the path correct to the nearest centimetre if the area is $400\,\text{m}^2$.

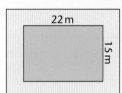

K KEY IDEAS

■ To calculate the roots of a quadratic equation which will not factorise by completing the square

- ■ multiply if necessary so the coefficient of x^2 is a perfect square
- ■ add a number to both sides so the left-hand side is a complete square
- ■ take the square root of both sides
- ■ solve the resulting linear equations.

■ The roots of the equation $ax^2 + bx + c = 0$ can be found using the formula

$$x = \frac{-b \pm \sqrt{b^2 - 4ac}}{2a}.$$

STAGE
10

Surface areas and complex shapes

8

You will learn about

- Solving problems involving surface areas and volumes of pyramids, cylinders, cones and spheres
- Solving problems involving complex shapes, including segments of circles and frustums of cones

You should already know

- How to find the circumference and area of a circle
- How to find the arc length and area of a sector
- How to find the area of a triangle
- How to find the volume of a prism, pyramid, cone or sphere
- How to rearrange formulae
- How to use Pythagoras' theorem and trigonometry

Surface areas

 A **ACTIVITY 1**

a) Make a sketch of a rectangular-based pyramid.
What do you need to know to work out its volume?

b) Sketch the net of the same pyramid.
What do you need to know to work out its surface area?
Can you work it out knowing the base measurements and the height?

Think of the label around a cylindrical can. It can be opened out flat to form a rectangle.

The length of the rectangle is the circumference of the can. Its width is the height of the can. The area of the rectangle is the curved surface area of the cylinder.

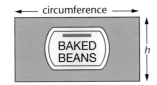

This gives the formula

Curved surface area of a cylinder = 2πrh

A ACTIVITY 2

a) Work with a partner.
Cut out a sector of a circle of radius 12 cm.
Each choose a different sector angle.

b) Calculate the arc length of your sector.
This will become the circumference of the base of a cone.

c) Stick the straight edges together to form a cone.

d) Work out the radius of the base of your cone.
Check your answer by measuring your cone.

e) The sector area has become the curved surface area of the cone.
Calculate this too.

C CHALLENGE 1

Replace 12 cm in Activity 2 with the slant height, l, of the cone.

Using a sector angle of θ, find the radius, r, of the base of the cone in terms of l and θ.

Then find the area of the sector in terms of r and l.

STAGE
10

The curved surface of a cone can be opened out to form the sector of a circle of radius *l*, where *l* is the slant height of the cone. The arc length of the sector is the circumference of the base of the cone.

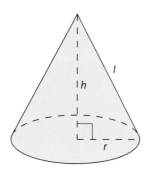
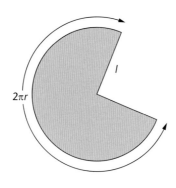

Curved surface area of a cone = $\pi r l$

You also need to known the formula for the surface area of a sphere of radius *r*.

Surface area of a sphere = $4\pi r^2$

EXAMPLE 1

This cone has a solid base.

a) Find the volume of the cone.

b) Find the total surface area of the cone.

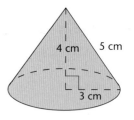

a) Volume = $\frac{1}{3}\pi r^2 h$

$= \frac{1}{3}\pi \times 3^2 \times 4$

$= 12\pi$

$= 37.7 \text{ cm}^3$ to 3 significant figures

b) Curved surface area = $\pi r l$

$= \pi \times 3 \times 5$

$= 15\pi$

Area of base = πr^2

$= \pi \times 3^2$

$= 9\pi$

Total surface area = $15\pi + 9\pi$

$= 24\pi$

$= 75.4 \text{ cm}^2$ to 3 significant figures

EXAMPLE 2

Calculate the curved surface area of this cone.

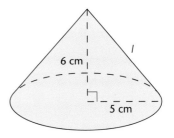

First the slant height *l* must be found.

Using Pythagoras,

$$l^2 = 5^2 + 6^2$$
$$= 61$$
$$l = \sqrt{61}$$

Curved surface area = $\pi r l$
$$= \pi \times 5 \times \sqrt{61}$$
$$= 123 \text{ cm}^2 \text{ to 3 significant figures}$$

EXERCISE 8.1

1 Calculate the curved surface area of a cylinder with these dimensions.
   a) Radius = 4·7 cm, height = 8·2 cm   **b)** Radius = 1·2 m, height = 2·5 m
   c) Radius = 3·5 cm, height = 4·6 cm   **d)** Radius = 2·7 cm, height = 3·4 cm
   e) Radius = 1·9 m, height = 1·6 m   **f)** Radius = 7·2 cm, height = 15·7 cm

2 Calculate the curved surface area of each of these cones.
   a)   **b)**   **c)**

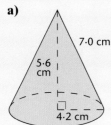   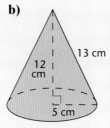   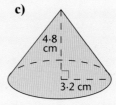

   d)   **e)**   **f)**

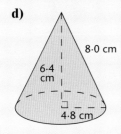   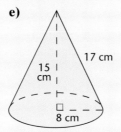   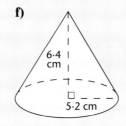

3 Calculate the surface area of a sphere of each of these radii.
   a) 5 cm   **b)** 6·2 cm
   c) 2 mm   **d)** 3 cm
   e) 4·7 cm   **f)** 7·8 mm

4 What is the total surface area of this cylinder?

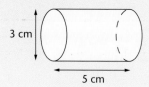

5 A sphere has a surface area of 47·6 cm². Calculate its radius.

6 A solid cone has a base radius of 4·5 cm and a height of 6·3 cm.
   Calculate its total surface area.

7 Calculate the total surface area of a solid cone with base radius 7·1 cm and slant
   height 9·7 cm.

8 A wastepaper bin is a cylinder with no lid. It is 50 cm high and the diameter of its
   base is 30 cm.
   The outside of the bin is painted white; the inside is painted black.
   Calculate the area that is painted black.

9 Calculate the total surface area of a solid hemisphere of radius 5·2 cm.

10 Calculate the slant height of a cone of base radius 5 cm and surface area 120 cm².

11 A solid cylinder of length 6·0 cm has a curved surface area of 1800 cm².
   Calculate its radius.

12 A cylinder has height 7·2 cm and base area 36 cm².
   Calculate its curved surface area.

13 The flat surface of a hemisphere has an area of 85 cm².
   Calculate the curved surface area.

14 A sphere has a surface area of 157·6 cm².
   Calculate its radius.

15 A cylinder 6·3 cm long has a curved surface area of 170 cm².
   Calculate the radius of the cylinder.

16 Calculate the base area of a cone with slant height 8·2 cm and curved surface area 126 cm².

17 A cone has slant height 7 cm and curved surface area 84 cm².
   Calculate the total surface area of the cone.

More complex problems

These may involve combining shapes you have met earlier. Or they may use topics you have met in other chapters, such as Pythagoras' theorem and trigonometry.
Be prepared for anything, and enjoy the problem-solving!

One shape you may not have met before is the **frustum** of a cone. This is the shape remaining when a solid cone has a smaller cone removed from it as shown in this diagram. The circle on the top of the frustum is in a plane parallel to the base.

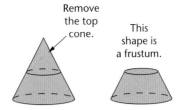

Remove the top cone.

This shape is a frustum.

> **Volume of frustrum = volume of whole cone – volume of missing cone**

EXAM TIP

When dealing with problems where you first have to work out how to solve them, follow these steps.

- Read the question carefully and plan.
  What do I know?
  What do I have to find?
  What methods can I apply?

- Look back when you have finished and ask 'Have I answered the question?'
  There may be one last step you have forgotten to do.

EXAMPLE 3

Calculate the area of the mauve segment.

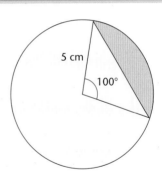

5 cm

100°

Area of segment = area of sector – area of triangle

Area of sector $= \dfrac{\theta}{360} \times \pi \times r^2$

$= \dfrac{100}{360} \times \pi \times 5^2$

$= 21 \cdot 816\ldots$ cm^2

Area of triangle $= \frac{1}{2}ab \sin C$

$= \frac{1}{2} \times 5^2 \times \sin 100°$

$= 12 \cdot 31\ldots$ cm^2

Area of segment $= 21 \cdot 816\ldots - 12 \cdot 31\ldots$

$= 9 \cdot 51$ cm^2 to 3 significant figures

EXAM TIP

Write down more figures than you need in the working, and round the final answer. Using the calculator memory means you do not have to re-key the figures.

EXAMPLE 4

Find the volume of the frustum remaining when a cone of height 8 cm is removed from a cone of height 12 cm and base radius 6 cm.

First, use similar triangles to find the base radius, r cm, of the cone which has been removed.

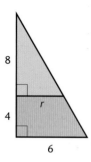

$$\frac{r}{8} = \frac{6}{12}$$

$$r = 4$$

Then find the volume of the frustum.

Volume of frustum = volume of whole cone − volume of missing cone

$$= \tfrac{1}{3}\pi R^2 H - \tfrac{1}{3}\pi r^2 h$$

$$= \tfrac{1}{3}\pi \times 6^2 \times 12 - \tfrac{1}{3}\pi \times 4^2 \times 8$$

$$= 318 \text{ cm}^2 \text{ to 3 significant figures}$$

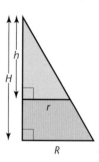

EXERCISE 8.2

1 A solid cylinder has a base radius of 3 cm.
Its volume is 95 cm³.
Calculate its curved surface area.

2 Calculate
   a) the length of the chord AB.
   b) the perimeter of the pink segment.

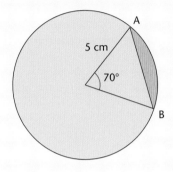

3 A cone of height 15 cm and base 9 cm has a cone of height 5 cm removed from its top as shown.

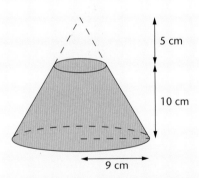

   a) What is the radius of the base of the top cone?
   b) Calculate the volume of the remaining frustum of the cone.

4 Show that the volume of this frustum is $\frac{7}{3}\pi r^2 h$.

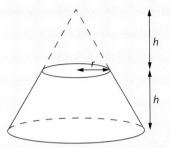

5 All the edges of this square-based pyramid are 5 cm.

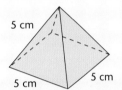

   Calculate
   a) its perpendicular height.
   b) its volume.

6 **a)** Show that the area of the green segment may be expressed as

$$r^2\left(\frac{50\pi}{360} - \frac{\sin 50°}{2}\right)$$

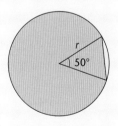

   b) Calculate the radius of the circle if the area of the green segment is 2 cm².

7 A cone has height 6 cm and volume 70 cm³.
   a) Calculate its base radius.
   b) Calculate its curved surface area.

8 A paintball sphere has a capacity of 1 litre.
   Calculate the surface area of the sphere.

   Hint: Capacity is the amount (volume) that a container can hold.

9 Calculate the area of the major segment (shaded yellow) in this diagram.

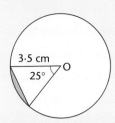

10 A piece of cheese is a prism whose cross-section is the sector of a circle with measurements as shown.

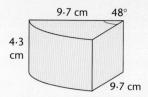

   Calculate the volume of the piece of cheese.

11 Calculate the area of the mauve segment.

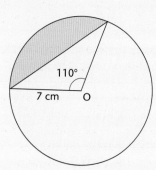

12 Calculate the perimeter of this segment of a circle of radius 5 cm.

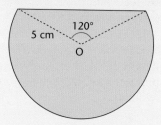

13 The bowl of this glass is part of a sphere.
   The radius of the sphere is 5 cm.
   The radius of the top of the glass is 3·7 cm.

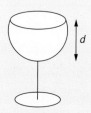

   Calculate the depth, d, of the glass.

14 The diagram shows the frustum of a cone.

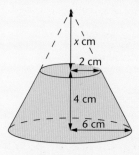

   a) Explain why $x + 4 = 3x$.
   b) Calculate the volume of the frustum.

15 A cone has a base of radius 5 cm and a perpendicular height of 12 cm.
   Calculate its curved surface area.

16 The perpendicular height of a cone is equal to its base radius.
The volume of the cone is $24\,\text{cm}^3$.
Calculate its perpendicular height.

17 The faces labelled A and B of this slice of cake are covered in chocolate.
The complete cake is a cylinder of radius 9 cm and depth 7 cm.

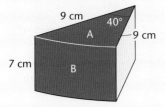

What area of the slice is covered in chocolate?

18 A spherical ball has a curved surface area of $120\,\text{cm}^2$.
Calculate its volume.

19 A lampshade is made from the frustum of a hollow cone.

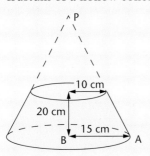

a) Show that the slant height, AP of the complete cone is $15\sqrt{17}\,\text{cm}$.

b) Calculate the curved surface area of the lampshade.

20 All the sloping edges of this square-based pyramid are 8 cm long.

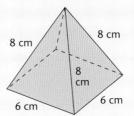

Calculate the volume of the pyramid.

21 A solid cone has a base of radius 6·9 cm and a height of 8·2 cm.
a) Calculate its volume.
b) Find its slant height and hence its total surface area.

22 The top of a flowerpot is a circle of radius 10 cm.
Its base is a circle of radius 8 cm.
The height of the flowerpot is 10 cm.

a) Show that the flowerpot is a frustum of an inverted cone of complete height 50 cm and base radius 10 cm.
b) Calculate how many litres of soil the flowerpot can contain.

23 A cone has a height of 20 cm and a base radius of 12 cm.
The top 15 cm of the cone is removed to leave a frustum of height 5 cm.
Find the volume of this frustum.

24 A pyramid has a square base of side 10·4 cm and its sloping edges are all 8·8 cm long.
Calculate the height and volume of the pyramid.

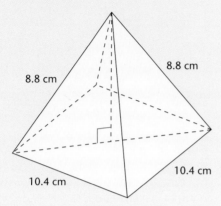

8.8 cm

8.8 cm

10.4 cm

10.4 cm

25 A sector with angle 300° and radius 12·5 cm is joined to form a hollow cone.
Showing your method clearly, calculate the volume of the cone.

CHALLENGE 2

The curved surface area of a particular cylinder of radius *r* cm and height *h* cm has the same surface area as a cube of side *r* cm.

Find *h* in terms of *r*.

KEY IDEAS

- Volume of cylinder = $\pi r^2 h$

  Curved surface area of cylinder = $2\pi r h$

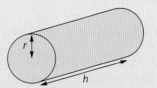

- Volume of pyramid = $\frac{1}{3}$ length $\times$ width $\times$ height

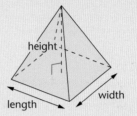

- Arc length of sector = $\dfrac{\theta}{360} \times 2\pi r$

  Area of sector = $\dfrac{\theta}{360} \times \pi r^2$

- Volume of cone = $\frac{1}{3}\pi r^2 h$

  Curved surface area of cone = $\pi r l$

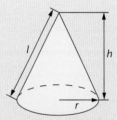

- Volume of sphere = $\frac{4}{3}\pi r^3$

  Surface area of sphere = $4\pi r^2$

- The frustum of a cone is the shape remaining when a solid cone has a smaller cone removed from it, as shown in this diagram.

  Volume of frustum =
  volume of whole cone – volume of missing cone

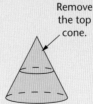

Remove the top cone.

This shape is a frustum.

- Area of minor segment =
  area of sector – area of triangle

STAGE
10

Revision exercise B1

1 The figures in the table show Colin's electricity bills (in £).

	1st quarter	2nd quarter	3rd quarter	4th quarter
2003	120·34	78·61	56·98	110·55
2004	126·92	75·03	55·09	120·81
2005	132·67	81·32	61·14	123·50
2006	143·84	79·89	70·83	125·16

a) Plot these figures on a graph.
b) Calculate the four-point moving averages and plot them on your graph.
c) What do you notice?
d) Predict the bills for the next four quarters.

2 The table gives the rainfall in millimetres for each month in Huangogo in Central Africa during a three-year period.

	J	F	M	A	M	J	J	A	S	O	N	D
2004	25	40	67	104	15	2	0	0	0	4	29	21
2005	30	38	80	116	10	0	0	0	6	7	36	19
2006	28	46	91	115	18	6	0	0	4	11	40	23

a) In Central Africa there is a rainy season and a dry season. Identify when these are.
b) Calculate a suitable moving average and comment on any trend.

3 The graph shows the temperature of a commercial oven used for baking large quantities of pastry at one time.

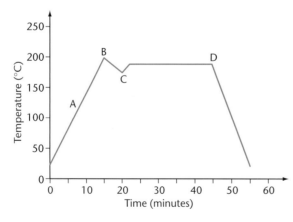

a) Describe what is happening at points A, B, C, and D.
b) Find the rate of warming up at A.

4 Join the midpoints of the sides of a square to form a quadrilateral.
Prove that this quadrilateral is a square.

5 PA and PB are tangents to the circle with centre O.
Prove that triangles AOP and BOP are congruent.

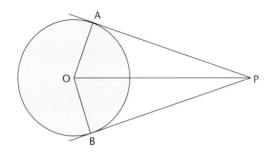

6 In the diagram for question **5**, let D be the point where AB crosses OP.
Prove that triangles PAD and PBD are congruent.

7 Solve these quadratic equations by completing the square.
Give your answers correct to 2 decimal places.
   a) $x^2 - 6x + 2 = 0$
   b) $4x^2 - 5x - 3 = 0$
   c) $3x^2 + 4x - 2 = 0$
   d) $x^2 + 13x - 27 = 0$
   e) $2x^2 + 10x - 19 = 0$

8 Write $y = x^2 - 5x + 4$ in the form $y = (x - a)^2 + b$.
Hence state the coordinates of the lowest point on the graph of $y = x^2 - 5x + 4$.

9 A rectangular lawn is three times as wide as it is long.
It has a path 1 m wide round three sides as shown in the diagram.

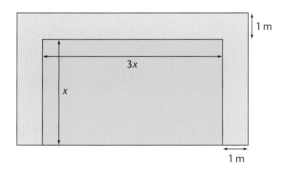

The area of the path, shaded brown, is equal to the area of the lawn.
   a) Explain why $3x^2 = 5x + 2$.
   b) Solve the equation $3x^2 - 5x - 2 = 0$ to find the dimensions of the lawn.

STAGE
10

10 Use the quadratic formula to solve these equations.
Give your answers correct to 3 decimal places.
a) $5x^2 - 8x + 1 = 0$
b) $x^2 - 7x - 2 = 0$
c) $6x^2 + 2x - 7 = 0$
d) $3x^2 + 5x - 10 = 0$
e) $5x^2 + 3x - 4 = 0$

11 O is the centre of the circle.
Calculate the area of the blue segment.

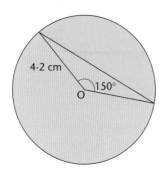

4·2 cm   150°   O

12 A solid cone and a solid cylinder both have base radius 6 cm.
The height of the cylinder is 4 cm.
The cone and the cylinder both have the same volume.
a) Find the height of the cone.
b) Calculate the curved surface area of the cylinder.

13 A sphere has volume $50 \, \text{cm}^3$.
Calculate its surface area.

14 A traffic bollard consists of a sphere on top of a cylinder.
The radii of the sphere and cylinder are each 12 cm.
The height of the cylinder is 90 cm.
One litre of black paint covers $4 \, \text{m}^2$.
How many of these bollards can be painted with 10 litres of paint?

90 cm

12 cm

Working with algebraic fractions

9

You will learn about

- Manipulating algebraic expressions, including those with fractions
- Solving equations, including those involving algebraic fractions

You should already know

- How to add, subtract, multiply, divide and simplify numerical fractions
- How to simplify and factorise algebraic expressions
- How to rearrange formulae

Algebraic fractions

A | ACTIVITY 1

a) Remind yourself how to add and subtract fractions.

(i) $\dfrac{2}{3} + \dfrac{5}{8}$ **(ii)** $\dfrac{3}{4} + \dfrac{7}{20}$

(iii) $\dfrac{9}{8} - \dfrac{2}{15}$ **(iv)** $\dfrac{x}{3} + \dfrac{5x}{8}$

(v) $\dfrac{5x}{9} - \dfrac{4x}{15}$

b) Write these fractions in their lowest terms.

(i) $\dfrac{5}{20}$ **(ii)** $\dfrac{16}{24}$

(iii) $\dfrac{3x}{9}$ **(iv)** $\dfrac{6}{27x}$

(v) $\dfrac{9a}{6a^2}$

STAGE
10

When an expression involving fractions needs to be simplified, all the fractions should be written with a common denominator.

EXAMPLE 1

Simplify $\dfrac{x+3}{2} - \dfrac{x-3}{3}$.

The common denominator is 6.

So $(x + 3)$ is multiplied by $6 \div 2 = 3$ and $(x - 3)$ is multiplied by $6 \div 3 = 2$.

$$\frac{x+3}{2} - \frac{x-3}{3} = \frac{3(x+3) - 2(x-3)}{6}$$
$$= \frac{3x + 9 - 2x + 6}{6}$$
$$= \frac{x + 15}{6}$$

> **EXAM TIP**
> Do not miss out the first step. Most errors occur because of expanding the brackets wrongly without writing them down.

If the denominators involve x the procedure is still the same.

EXAMPLE 2

Simplify $\dfrac{3}{x+1} - \dfrac{2}{x}$.

The common denominator is $x(x + 1)$.

So 3 is multiplied by $x(x + 1) \div (x + 1) = x$ and 2 is multiplied by $x(x + 1) \div x = (x + 1)$.

$$\frac{3}{x+1} - \frac{2}{x} = \frac{3x - 2(x+1)}{x(x+1)}$$
$$= \frac{3x - 2x - 2}{x(x+1)}$$
$$= \frac{x - 2}{x(x+1)}$$

> **EXAM TIP**
> Errors often occur through cancelling individual terms. Only factors, which can be individual numbers, letters or brackets, can be cancelled.

EXAMPLE 3

Simplify $\dfrac{x}{x+3} - \dfrac{x-2}{x} + \dfrac{2}{5}$.

The common denominator is $5x(x+3)$.

Multiply x by $5x$, $(x-2)$ by $5(x+3)$ and 2 by $x(x+3)$.

$$\frac{x}{x+3} - \frac{x-2}{x} + \frac{2}{5} = \frac{5x^2 - 5(x+3)(x-2) + 2x(x+3)}{5x(x+3)}$$

$$= \frac{5x^2 - 5(x^2 + x - 6) + 2x^2 + 6x}{5x(x+3)}$$

$$= \frac{5x^2 - 5x^2 - 5x + 30 + 2x^2 + 6x}{5x(x+3)}$$

$$= \frac{2x^2 + x + 30}{5x(x+3)}$$

EXERCISE 9.1

Simplify these.

1 $\dfrac{x}{2} + \dfrac{3x}{5}$

2 $\dfrac{2x}{3} - \dfrac{3x}{5}$

3 $\dfrac{x+1}{3} - \dfrac{2x-1}{2}$

4 $\dfrac{x-1}{2} - \dfrac{x-3}{5}$

5 $\dfrac{x-3}{5} + \dfrac{2x}{3} - \dfrac{3x-2}{10}$

6 $\dfrac{2x-1}{6} + \dfrac{3x}{4} - \dfrac{x-2}{12}$

7 $\dfrac{1}{x} + \dfrac{2}{x-1}$

8 $\dfrac{3}{x} + \dfrac{2}{x+1}$

9 $\dfrac{3}{2x} - \dfrac{1}{x+2}$

10 $\dfrac{5}{6x} - \dfrac{1}{2x+1}$

11 $\dfrac{2}{x+1} + \dfrac{3}{x-1}$

12 $\dfrac{3}{x+2} + \dfrac{5}{x-1}$

13 $\dfrac{2x}{3x+1} - \dfrac{5}{x+3}$

14 $\dfrac{2x}{x+1} - \dfrac{x-1}{x+3}$

15 $\dfrac{x+1}{x-1} + \dfrac{3x-1}{x+2}$

16 $\dfrac{2x}{x-1} - \dfrac{3x+2}{x+2}$

STAGE
10

EXERCISE 9.1 continued

17 $\dfrac{x}{x+1} - \dfrac{3}{5} + \dfrac{x-2}{x}$

18 $\dfrac{x}{x+1} + \dfrac{3}{5} - \dfrac{x+3}{x}$

19 $\dfrac{2x}{x-3} + \dfrac{x-1}{x+2} - \dfrac{4}{9}$

20 $\dfrac{2}{x-1} - \dfrac{3}{x+2} - \dfrac{1}{x+3}$

21 $\dfrac{2}{2x+1} + \dfrac{3x+5}{x+2}$

22 $\dfrac{4x+17}{x+3} - \dfrac{2x-15}{x-3}$

Solving harder equations

You need to be able to solve equations involving brackets and algebraic fractions. These may be linear or quadratic.

EXAMPLE 4

> **EXAM TIP**
>
> In equations with algebraic fractions multiply through by the common denominator. This removes the fractions.

Solve $\dfrac{3}{x+1} - \dfrac{2}{x} = \dfrac{1}{x-2}$.

$$\frac{3}{x+1} - \frac{2}{x} = \frac{1}{x-2}$$

$3x(x-2) - 2(x+1)(x-2) = x(x+1)$ Multiply by $x(x+1)(x-2)$. Make sure you multiply every expression on both sides.

$3x^2 - 6x - 2(x^2 - x - 2) = x^2 + x$

$3x^2 - 6x - 2x^2 + 2x + 4 - x^2 - x = 0$ Expand the bracket and collect on one side.

$^-5x + 4 = 0$ Collect like terms.

$5x = 4$ Rearrange and change signs.

$x = \dfrac{4}{5}$

STAGE

10

9

EXAMPLE 5

Solve $(x - 1)^2 = 7 - x$.

$(x - 1)^2 = 7 - x$ The first thing to do is multiply out the bracket.

$x^2 - 2x + 1 = 7 - x$ Remember that $(x - 1)^2 = (x - 1)(x - 1)$.

$x^2 - x - 6 = 0$ Collect all the terms on one side.

$(x - 3)(x + 2) = 0$ Factorise the expression.

$x = 3$ or $^-2$

EXAMPLE 6

Solve $\dfrac{5}{x} = \dfrac{x}{x + 10}$.

$$\dfrac{5}{x} = \dfrac{x}{x + 10}$$

$5(x + 10) = x^2$ Multiply through by $x(x + 10)$.

$5x + 50 = x^2$ Expand the bracket.

$^-x^2 + 5x + 50 = 0$ Collect all the terms on one side.

$x^2 - 5x - 50 = 0$ Change all the signs.

$(x - 10)(x + 5) = 0$ Factorise the expression.

$x = 10$ or $^-5$

EXERCISE 9.2

Solve these.

1 $\dfrac{2x}{3} - \dfrac{3x}{5} = \dfrac{1}{3}$

2 $\dfrac{2x}{5} - \dfrac{x}{4} = \dfrac{3}{10}$

3 $\dfrac{x-1}{2} - \dfrac{x-3}{5} = 1$

4 $\dfrac{2x-1}{2} - \dfrac{x-3}{3} = \dfrac{5}{2}$

5 $\dfrac{2x-1}{6} + \dfrac{3x}{4} = \dfrac{x-2}{12}$

6 $\dfrac{x-1}{3} + \dfrac{2x}{5} = \dfrac{3x+1}{5}$

7 $\dfrac{3}{x} - \dfrac{2}{x+1} = 0$

8 $\dfrac{5}{x} - \dfrac{2}{x-3} = 0$

9 $\dfrac{5}{6x} - \dfrac{1}{x+1} = \dfrac{1}{3x}$

10 $\dfrac{4}{x-2} - \dfrac{1}{x+1} = \dfrac{3}{x}$

11 $x(x+2) = 2(x+2)$

12 $(x+4)(x+2) + x + 4 = 0$

13 $2x(x-2) = x^2 + 5$

14 $(x-5)(x+3) = x - 5$

15 $4x = \dfrac{3}{x} - 1$

16 $4x + \dfrac{3}{x} = 7$

17 $2x^2 - \dfrac{x}{3} = 5$

18 $2x + \dfrac{4}{x} = 9$

19 $\dfrac{1}{x-1} - \dfrac{3}{x+2} = \dfrac{1}{4}$

20 $\dfrac{2x}{3x+1} - \dfrac{5}{x+3} = 0$

21 $\dfrac{2x}{x-3} - \dfrac{x}{x-2} = 3$

22 $\dfrac{2}{x} + \dfrac{1}{x+1} = 5$

23 $\dfrac{x}{x-2} - 2x = 3$

24 $\dfrac{2x}{2x-5} + \dfrac{x-1}{3x} = 2$

STAGE 10

K KEY IDEAS

- When adding or subtracting fractions, put them over a common denominator.

- When cancelling algebraic fractions, factorise if necessary. Only cancel factors.

- When equations involve fractions, multiply through by the common denominator to remove the fractions.

Vectors 10

Column vectors

If a **vector** is drawn on a grid then it can be described by a column vector $\begin{pmatrix} x \\ y \end{pmatrix}$, where x is the length across to the right and y is the length upwards.

In the diagram $\mathbf{a} = \begin{pmatrix} 2 \\ 1 \end{pmatrix}$ $\mathbf{b} = \begin{pmatrix} 3 \\ -2 \end{pmatrix}$ $\mathbf{c} = \begin{pmatrix} -3 \\ -1 \end{pmatrix}$

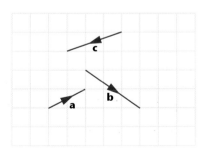

STAGE
10

EXAM TIP

Column vectors must be columns. If you write them down as coordinates they will be marked as wrong.

EXAMPLE 1

Write down the column vectors for

$\overrightarrow{AB}, \overrightarrow{BC}, \overrightarrow{CD}, \overrightarrow{AD}, \overrightarrow{BD}, \overrightarrow{DC}.$

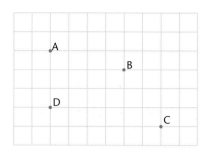

$\overrightarrow{AB} = \begin{pmatrix} 4 \\ -1 \end{pmatrix}$, $\overrightarrow{BC} = \begin{pmatrix} 2 \\ -3 \end{pmatrix}$, $\overrightarrow{CD} = \begin{pmatrix} -6 \\ 1 \end{pmatrix}$,

$\overrightarrow{AD} = \begin{pmatrix} 0 \\ -3 \end{pmatrix}$, $\overrightarrow{BD} = \begin{pmatrix} -4 \\ -2 \end{pmatrix}$, $\overrightarrow{DC} = \begin{pmatrix} 6 \\ -1 \end{pmatrix}$

EXAM TIP

In examinations, candidates often make an error of 1 when working out the values for the vector. Take care with the counting.

 ### ACTIVITY 1

a) Draw these column vectors on squared paper.

$\begin{pmatrix} 1 \\ 3 \end{pmatrix} \begin{pmatrix} 2 \\ 6 \end{pmatrix} \begin{pmatrix} 3 \\ 9 \end{pmatrix}$

What do you notice?

b) Repeat with these vectors.

$\begin{pmatrix} 3 \\ 2 \end{pmatrix} \begin{pmatrix} 6 \\ 4 \end{pmatrix} \begin{pmatrix} 9 \\ 6 \end{pmatrix}$

You have met column vectors before, when you studied translation.

▌▌EXAMPLE 2

Find the column vector that maps

a) (2, 1) on to (5, 6).

b) (⁻1, 6) on to (5, 2).

a) The x-coordinate has changed from 2 to 5 so the increase is 3.
The y-coordinate has changed from 1 to 6 so the increase is 5.

The vector is $\begin{pmatrix} 3 \\ 5 \end{pmatrix}$.

b) The x-coordinate has changed from ⁻1 to 5 so the increase is 6.
The y-coordinate has changed from 6 to 2 so the decrease is 4.

The vector is $\begin{pmatrix} 6 \\ -4 \end{pmatrix}$.

EXAM TIP

As you can see, it is not
necessary to plot the points to
do this question. However,
you may prefer to do so.

General vectors

A vector has both length and direction but can be in any position. The vector going from A to B
can be labelled $\overrightarrow{AB}$ or it can be given a letter **a**, in bold type. When handwritten, put a wavy line
underneath, a̰ for example.

All the four lines drawn below are of equal length and go in the same direction, and they can all
be called **a**.

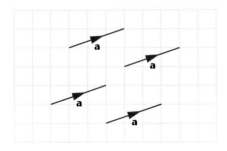

Look at this diagram.

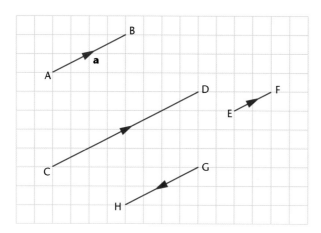

$\overrightarrow{AB}$ = **a**.

The line $\overrightarrow{CD}$ is parallel to $\overrightarrow{AB}$ and twice as long so $\overrightarrow{CD}$ = 2**a**.

EF is parallel to AB and half the length so $\overrightarrow{EF} = \frac{1}{2}$**a**.

GH is parallel and equal in length to BA (opposite direction to AB) so $\overrightarrow{GH}$ = ⁻**a**.

▐ EXAMPLE 3

For the diagram below write down the vectors $\overrightarrow{CD}$, $\overrightarrow{EF}$, $\overrightarrow{GH}$ and $\overrightarrow{PQ}$ in terms of **a**.

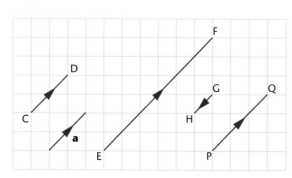

$\overrightarrow{CD}$ = **a**, $\overrightarrow{EF}$ = 3**a**, $\overrightarrow{GH} = \frac{-1}{2}$**a**, $\overrightarrow{PQ} = \frac{3}{2}$**a**

EXAMPLE 4

ABCD is a rectangle and E, F, G, H are the midpoints of the sides.

AB = **a** and AD = **b**.

Write the vectors $\overrightarrow{BC}$, $\overrightarrow{CD}$, $\overrightarrow{AE}$, $\overrightarrow{AH}$, $\overrightarrow{EG}$, $\overrightarrow{CF}$ and $\overrightarrow{FH}$ in terms of **a** or **b**.

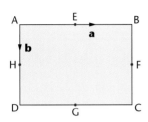

$\overrightarrow{BC} = $ **b**, $\overrightarrow{CD} = $ ⁻**a**, $\overrightarrow{AE} = \frac{1}{2}$**a**, $\overrightarrow{AH} = \frac{1}{2}$**b**, $\overrightarrow{EG} = $ **b**, $\overrightarrow{CF} = \frac{-1}{2}$**b**, $\overrightarrow{FH} = $ ⁻**a**

EXERCISE 10.1

1 Write down the column vectors for $\overrightarrow{AB}$, $\overrightarrow{CD}$, $\overrightarrow{CB}$, $\overrightarrow{AD}$ and $\overrightarrow{CA}$.

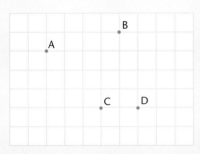

2 Write down the column vectors for $\overrightarrow{EF}$, $\overrightarrow{GH}$, $\overrightarrow{EH}$, $\overrightarrow{GF}$ and $\overrightarrow{FH}$.

3 Write down the column vectors for $\overrightarrow{AB}$, $\overrightarrow{CD}$, $\overrightarrow{CB}$, $\overrightarrow{AD}$ and $\overrightarrow{CA}$.

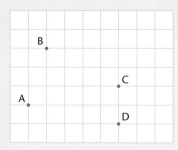

4 Write down the column vectors for $\overrightarrow{AB}$, $\overrightarrow{CD}$, $\overrightarrow{CB}$, $\overrightarrow{AD}$ and $\overrightarrow{CA}$.

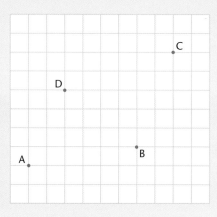

5 Find the column vector that maps
   a) $(1, 2)$ on to $(1, 4)$.
   b) $(2, 3)$ on to $(^-2, 3)$.
   c) $(1, 0)$ on to $(^-1, 3)$.
   d) $(4, 2)$ on to $(5, 9)$.
   e) $(^-3, 2)$ on to $(5, ^-4)$.
   f) $(6, 1)$ on to $(0, 5)$.

6 For the diagram below, write down the vectors $\overrightarrow{AB}$, $\overrightarrow{CD}$, $\overrightarrow{EF}$, $\overrightarrow{GH}$, $\overrightarrow{PQ}$ and $\overrightarrow{RS}$ in terms of **a**.

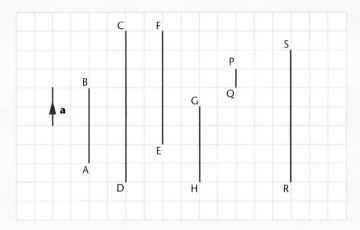

7 Copy and complete this table.

	Original point	Vector	New point
a)	(1, 2)	$\begin{pmatrix} 3 \\ 2 \end{pmatrix}$	
b)	(2, 3)	$\begin{pmatrix} 4 \\ 1 \end{pmatrix}$	
c)	(1, 0)	$\begin{pmatrix} {}^-3 \\ 2 \end{pmatrix}$	
d)	(4, 2)	$\begin{pmatrix} 0 \\ {}^-3 \end{pmatrix}$	
e)	($^-$3, 2)	$\begin{pmatrix} {}^-5 \\ {}^-2 \end{pmatrix}$	
f)	(6, 1)	$\begin{pmatrix} {}^-6 \\ {}^-1 \end{pmatrix}$	
g)	(2, 4)	$\begin{pmatrix} 3 \\ 2 \end{pmatrix}$	
h)	(3, 2)	$\begin{pmatrix} 5 \\ {}^-1 \end{pmatrix}$	
i)	(6, 3)	$\begin{pmatrix} {}^-4 \\ {}^-2 \end{pmatrix}$	
j)	($^-$1, 5)	$\begin{pmatrix} 3 \\ {}^-4 \end{pmatrix}$	
k)	($^-$4, $^-$3)	$\begin{pmatrix} 5 \\ {}^-2 \end{pmatrix}$	
l)	(6, $^-$2)	$\begin{pmatrix} 2 \\ {}^-4 \end{pmatrix}$	

STAGE

10

8 For the diagram below, write down the vectors $\overrightarrow{AB}$, $\overrightarrow{CD}$, $\overrightarrow{EF}$, $\overrightarrow{GH}$, $\overrightarrow{PQ}$ and $\overrightarrow{RS}$ in terms of **a**.

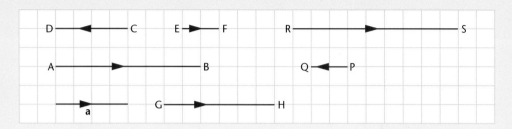

9 For the diagram below, write down the vectors $\overrightarrow{AB}$, $\overrightarrow{CD}$, $\overrightarrow{EF}$, $\overrightarrow{GH}$, $\overrightarrow{PQ}$ and $\overrightarrow{RS}$ in terms of **a** or **b**.

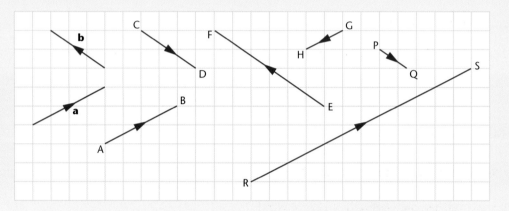

10 For the diagram below, write down the vectors $\overrightarrow{AB}$, $\overrightarrow{CD}$, $\overrightarrow{EF}$, $\overrightarrow{GH}$, $\overrightarrow{PQ}$ and $\overrightarrow{RS}$ in terms of **a** or **b**.

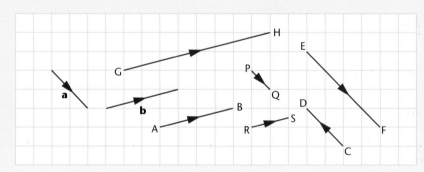

EXERCISE 10.1 continued

11 ABCD is a parallelogram. E, F, G, H are the midpoints of the sides.

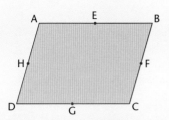

$\overrightarrow{AE}$ = **a** and $\overrightarrow{AH}$ = **b**.

Write down the vectors for $\overrightarrow{AB}$, $\overrightarrow{CD}$, $\overrightarrow{EB}$, $\overrightarrow{GD}$, $\overrightarrow{HF}$ and $\overrightarrow{FC}$ in terms of **a** or **b**.

12 ABCD is a square. E, F, G, H are the midpoints of sides AB, BC, CD, DA respectively.

$\overrightarrow{AB}$ = **a** and $\overrightarrow{AD}$ = **b**.

Write down the vectors $\overrightarrow{BC}$, $\overrightarrow{CD}$, $\overrightarrow{EB}$, $\overrightarrow{HD}$, $\overrightarrow{HF}$ and $\overrightarrow{FB}$ in terms of **a** or **b**.

Multiplying a vector by a scalar

In the diagram you can see that $\mathbf{a} = \begin{pmatrix} 2 \\ 1 \end{pmatrix}$.

You can also see that $\overrightarrow{AB}$ = 2**a** and $\overrightarrow{CD}$ = 3**a**.

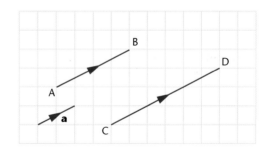

$\overrightarrow{AB} = \begin{pmatrix} 4 \\ 2 \end{pmatrix} = 2 \times \begin{pmatrix} 2 \\ 1 \end{pmatrix} = 2\mathbf{a}$

$\overrightarrow{CD} = \begin{pmatrix} 6 \\ 3 \end{pmatrix} = 3 \times \begin{pmatrix} 2 \\ 1 \end{pmatrix} = 3\mathbf{a}$

This shows that $k \times \begin{pmatrix} a \\ b \end{pmatrix} = \begin{pmatrix} ka \\ kb \end{pmatrix}$.

A quantity that has magnitude but not direction is called a **scalar**.

Multiplying a vector by a scalar produces a vector in the same direction but longer by a factor equal to the scalar.

If you know that $\overrightarrow{AB} = k\overrightarrow{CD}$, you can conclude that

- $\overrightarrow{AB}$ is parallel to $\overrightarrow{CD}$.
- $\overrightarrow{AB}$ is k times the length of $\overrightarrow{CD}$.

If you know that $\overrightarrow{AB} = k\overrightarrow{AC}$ and there is a common point A, you can conclude that

- A, B and C are in a straight line.
- $\overrightarrow{AB}$ is k times the length of $\overrightarrow{AC}$.

Addition and subtraction of column vectors

In the diagram, $\mathbf{a} = \begin{pmatrix} 3 \\ 1 \end{pmatrix}$ and $\mathbf{b} = \begin{pmatrix} 2 \\ -3 \end{pmatrix}$.

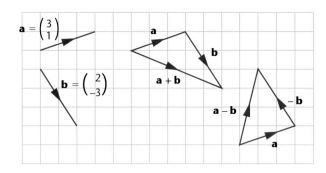

You can see that

$$\mathbf{a} + \mathbf{b} = \begin{pmatrix} 3 \\ 1 \end{pmatrix} + \begin{pmatrix} 2 \\ -3 \end{pmatrix} = \begin{pmatrix} 5 \\ -2 \end{pmatrix}$$

and $\mathbf{a} - \mathbf{b} = \mathbf{a} + (^-\mathbf{b}) = \begin{pmatrix} 3 \\ 1 \end{pmatrix} - \begin{pmatrix} 2 \\ -3 \end{pmatrix} = \begin{pmatrix} 1 \\ 4 \end{pmatrix}$.

So, to add or subtract column vectors, you simply add or subtract the components.

$$\begin{pmatrix} a \\ b \end{pmatrix} + \begin{pmatrix} c \\ d \end{pmatrix} = \begin{pmatrix} a + c \\ b + d \end{pmatrix} \quad \text{and} \quad \begin{pmatrix} a \\ b \end{pmatrix} - \begin{pmatrix} c \\ d \end{pmatrix} = \begin{pmatrix} a - c \\ b - d \end{pmatrix}$$

STAGE
10

Vectors

10

264

EXAMPLE 5

Given $\mathbf{a} = \begin{pmatrix} 3 \\ 1 \end{pmatrix}$, $\mathbf{b} = \begin{pmatrix} 1 \\ 3 \end{pmatrix}$ and $\mathbf{c} = \begin{pmatrix} -2 \\ 1 \end{pmatrix}$, work out these.

a) $2\mathbf{a}$ b) $\mathbf{a} + 2\mathbf{b}$ c) $\mathbf{a} - \mathbf{b} + \mathbf{c}$

d) $2\mathbf{a} + \mathbf{b} - \mathbf{c}$ e) $\frac{1}{2}\mathbf{a}$

a) $2\mathbf{a} = 2 \times \begin{pmatrix} 3 \\ 1 \end{pmatrix}$

$= \begin{pmatrix} 6 \\ 2 \end{pmatrix}$

b) $\mathbf{a} + 2\mathbf{b} = \begin{pmatrix} 3 \\ 1 \end{pmatrix} + 2 \times \begin{pmatrix} 1 \\ 3 \end{pmatrix}$

$= \begin{pmatrix} 3 \\ 1 \end{pmatrix} + \begin{pmatrix} 2 \\ 6 \end{pmatrix}$

$= \begin{pmatrix} 5 \\ 7 \end{pmatrix}$

c) $\mathbf{a} - \mathbf{b} + \mathbf{c} = \begin{pmatrix} 3 \\ 1 \end{pmatrix} - \begin{pmatrix} 1 \\ 3 \end{pmatrix} + \begin{pmatrix} -2 \\ 1 \end{pmatrix}$

$= \begin{pmatrix} 0 \\ -1 \end{pmatrix}$

d) $2\mathbf{a} + \mathbf{b} - \mathbf{c} = 2 \times \begin{pmatrix} 3 \\ 1 \end{pmatrix} + \begin{pmatrix} 1 \\ 3 \end{pmatrix} - \begin{pmatrix} -2 \\ 1 \end{pmatrix}$

$= \begin{pmatrix} 6 \\ 2 \end{pmatrix} + \begin{pmatrix} 1 \\ 3 \end{pmatrix} - \begin{pmatrix} -2 \\ 1 \end{pmatrix}$

$= \begin{pmatrix} 9 \\ 4 \end{pmatrix}$

e) $\frac{1}{2}\mathbf{a} = \frac{1}{2} \times \begin{pmatrix} 3 \\ 1 \end{pmatrix}$

$= \begin{pmatrix} 1.5 \\ 0.5 \end{pmatrix}$

EXAM TIP
When adding and subtracting column vectors, be very careful with the signs as most errors are made in that way.

STAGE
10

EXERCISE 10.2

1 Work out these.

a) $2 \times \begin{pmatrix} 2 \\ 3 \end{pmatrix}$

b) $\begin{pmatrix} 6 \\ 2 \end{pmatrix} + \begin{pmatrix} 3 \\ 1 \end{pmatrix}$

c) $\frac{1}{2}\begin{pmatrix} 4 \\ 6 \end{pmatrix}$

d) $\begin{pmatrix} 3 \\ 1 \end{pmatrix} - \begin{pmatrix} 2 \\ 1 \end{pmatrix}$

e) $\begin{pmatrix} 3 \\ 4 \end{pmatrix} + 2 \times \begin{pmatrix} 1 \\ 4 \end{pmatrix}$

2 Work out these.

a) $2 \times \begin{pmatrix} ^-3 \\ 0 \end{pmatrix}$

b) $\begin{pmatrix} 3 \\ 1 \end{pmatrix} - \begin{pmatrix} 4 \\ 3 \end{pmatrix}$

c) $\frac{1}{2}\begin{pmatrix} 1 \\ -3 \end{pmatrix}$

d) $\begin{pmatrix} 2 \\ -1 \end{pmatrix} + 2 \times \begin{pmatrix} 2 \\ 1 \end{pmatrix}$

e) $\frac{1}{2}\begin{pmatrix} 1 \\ 4 \end{pmatrix} - \frac{1}{4}\begin{pmatrix} 2 \\ 4 \end{pmatrix}$

3 Work out these.

a) $3 \times \begin{pmatrix} 1 \\ 4 \end{pmatrix}$

b) $\begin{pmatrix} 3 \\ 4 \end{pmatrix} + \begin{pmatrix} 5 \\ 8 \end{pmatrix}$

c) $\frac{1}{2}\begin{pmatrix} 8 \\ 10 \end{pmatrix}$

d) $2 \times \begin{pmatrix} 5 \\ 4 \end{pmatrix} - \begin{pmatrix} 3 \\ 4 \end{pmatrix}$

e) $2 \times \begin{pmatrix} 1 \\ 4 \end{pmatrix} + 5 \times \begin{pmatrix} 1 \\ 2 \end{pmatrix}$

4 Work out these.

a) $2 \times \begin{pmatrix} ^-1 \\ 0 \end{pmatrix}$

b) $\begin{pmatrix} 1 \\ 6 \end{pmatrix} - \begin{pmatrix} 7 \\ 3 \end{pmatrix}$

c) $\frac{1}{2}\begin{pmatrix} ^-2 \\ 4 \end{pmatrix}$

d) $\begin{pmatrix} 1 \\ -4 \end{pmatrix} - 2 \times \begin{pmatrix} 2 \\ 3 \end{pmatrix}$

e) $\frac{1}{2}\begin{pmatrix} 2 \\ 6 \end{pmatrix} - \frac{1}{2}\begin{pmatrix} 3 \\ -5 \end{pmatrix}$

5 Given that $\mathbf{a} = \begin{pmatrix} 6 \\ 3 \end{pmatrix}$, work out these.

a) $2\mathbf{a}$

b) $^-\mathbf{a}$

c) $4\mathbf{a}$

d) $\frac{1}{2}\mathbf{a}$

e) $\frac{^-1}{3}\mathbf{a}$

6 Given that $\mathbf{a} = \begin{pmatrix} 1 \\ 3 \end{pmatrix}$ and $\mathbf{b} = \begin{pmatrix} 3 \\ 4 \end{pmatrix}$, work out these.

a) $3\mathbf{a}$

b) $\mathbf{a} + \mathbf{b}$

c) $\mathbf{b} - \mathbf{a}$

d) $2\mathbf{a} + \mathbf{b}$

e) $3\mathbf{a} - 2\mathbf{b}$

7 Given that $\mathbf{a} = \begin{pmatrix} 2 \\ 3 \end{pmatrix}$, $\mathbf{b} = \begin{pmatrix} ^-3 \\ 4 \end{pmatrix}$ and $\mathbf{c} = \begin{pmatrix} ^-1 \\ -3 \end{pmatrix}$, work out these.

a) $3\mathbf{c}$

b) $4\mathbf{c} - 2\mathbf{b}$

c) $\mathbf{a} - \mathbf{b} + \mathbf{c}$

d) $2\mathbf{a} + 3\mathbf{b} + 2\mathbf{c}$

e) $\frac{1}{2}\mathbf{a} - \mathbf{b} - \frac{1}{2}\mathbf{c}$

8 Given that $\mathbf{p} = \begin{pmatrix} 5 \\ 8 \end{pmatrix}$, work out these.

a) $4\mathbf{p}$

b) $^-2\mathbf{p}$

c) $\frac{1}{2}\mathbf{p}$

d) $9\mathbf{p}$

e) $\frac{2}{5}\mathbf{p}$

9 Given that $\mathbf{p} = \begin{pmatrix} 4 \\ 1 \end{pmatrix}$ and $\mathbf{q} = \begin{pmatrix} 5 \\ 3 \end{pmatrix}$, work out these.

a) $2\mathbf{p}$

b) $\mathbf{p} + \mathbf{q}$

c) $\mathbf{q} - \mathbf{p}$

d) $2\mathbf{p} + \mathbf{q}$

e) $3\mathbf{q} - 2\mathbf{p}$

10 Given that $\mathbf{a} = \begin{pmatrix} ^-2 \\ 4 \end{pmatrix}$, $\mathbf{b} = \begin{pmatrix} 3 \\ 5 \end{pmatrix}$ and $\mathbf{c} = \begin{pmatrix} ^-2 \\ -3 \end{pmatrix}$, work out these.

a) $3\mathbf{c}$

b) $3\mathbf{c} + 2\mathbf{b}$

c) $\mathbf{a} - \mathbf{b} + \mathbf{c}$

d) $\mathbf{a} + 4\mathbf{b} - 2\mathbf{c}$

e) $\frac{1}{2}\mathbf{a} + \mathbf{b} - \frac{1}{2}\mathbf{c}$

Vector geometry

You have already seen results like $\overrightarrow{PR} = \overrightarrow{PQ} + \overrightarrow{QR}$.

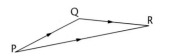

In the diagram, $\overrightarrow{OA} = \mathbf{a}$, $\overrightarrow{OB} = \mathbf{b}$ and OACB is a parallelogram.

AC is parallel and equal to OB so $\overrightarrow{AC} = \mathbf{b}$.

$\overrightarrow{OC} = \overrightarrow{OA} + \overrightarrow{AC} = \mathbf{a} + \mathbf{b}$.

$\overrightarrow{OC}$ is known as the **resultant** of $\mathbf{a}$ and $\mathbf{b}$.

You can also see in the diagram that $\overrightarrow{OC} = \overrightarrow{OB} + \overrightarrow{BC} = \mathbf{b} + \mathbf{a}$.

This shows that $\mathbf{a} + \mathbf{b} = \mathbf{b} + \mathbf{a}$. The vectors can be added in either order.

To do subtraction you use the fact that $\mathbf{p} - \mathbf{q} = \mathbf{p} + (^-\mathbf{q})$.

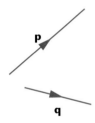
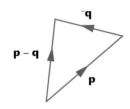

A ACTIVITY 2

Write down five different routes from A to E. You do not have to use all the points.

For each route, add together the vectors you have used.

What do you notice?

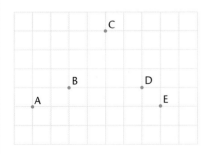

The resultant vector is the same, no matter which way you go.

This is a very important rule which, together with your knowledge of how to multiply a vector by a scalar, is used to find vectors in geometrical figures.

EXAMPLE 6

In the triangle ABC, $\overrightarrow{AB}$ = **p** and $\overrightarrow{AC}$ = **q** and D is the midpoint of BC.

Work out these vectors.

a) $\overrightarrow{BC}$

b) $\overrightarrow{BD}$

c) $\overrightarrow{AD}$

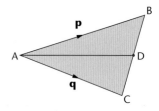

a) $\overrightarrow{BC} = \overrightarrow{BA} + \overrightarrow{AC}$
$\quad\quad = ^-\mathbf{p} + \mathbf{q}$
$\quad\quad = \mathbf{q} - \mathbf{p}$

b) $\overrightarrow{BD} = \frac{1}{2}\overrightarrow{BC}$
$\quad\quad = \frac{1}{2}(\mathbf{q} - \mathbf{p})$

c) $\overrightarrow{AD} = \overrightarrow{AB} + \overrightarrow{BD}$
$\quad\quad = \mathbf{p} + \frac{1}{2}(\mathbf{q} - \mathbf{p})$
$\quad\quad = \mathbf{p} + \frac{1}{2}\mathbf{q} - \frac{1}{2}\mathbf{p}$
$\quad\quad = \frac{1}{2}\mathbf{p} + \frac{1}{2}\mathbf{q}$
$\quad\quad = \frac{1}{2}(\mathbf{p} + \mathbf{q})$

EXAM TIP

If you were only asked to find $\overrightarrow{AD}$ directly, it would be an example of a multi-step question and you would need to work out which other vectors to find.

EXAMPLE 7

In this diagram, OC = 2 × OA and OD = 2 × OB.

$\overrightarrow{OA}$ = **a** and $\overrightarrow{OB}$ = **b**.

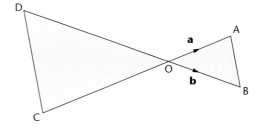

a) Work out these vectors.

(i) $\overrightarrow{OC}$

(ii) $\overrightarrow{OD}$

(iii) $\overrightarrow{AB}$

(iv) $\overrightarrow{DC}$

b) What does this show about the lines AB and DC?

a) (i) OC is on the same line as OA, in the opposite direction and twice as long.

$\overrightarrow{OC}$ = ⁻2 × $\overrightarrow{OA}$ = ⁻2**a**

(ii) By the same reasoning as above, $\overrightarrow{OD}$ = ⁻2 × $\overrightarrow{OB}$ = ⁻2**b**

(iii) $\overrightarrow{AB}$ = $\overrightarrow{AO}$ + $\overrightarrow{OB}$ = ⁻**a** + **b** = **b** − **a**

(iv) $\overrightarrow{DC}$ = $\overrightarrow{DO}$ + $\overrightarrow{OC}$ = 2**b** − 2**a** = 2(**b** − **a**)

b) The vector for DC is twice the vector for AB.

So AB and DC are parallel and DC is twice as long as AB.

EXERCISE 10.3

1 On a square grid with x and y from 0 to 8, plot A(1, 3) and B(3, 5).

    a) Write down $\overrightarrow{AB}$ as a column vector.

    b) Mark any two points as C and D and work out these.

      (i) $\overrightarrow{AC}$ + $\overrightarrow{CB}$

      (ii) $\overrightarrow{AD}$ + $\overrightarrow{DB}$

      (iii) $\overrightarrow{AC}$ + $\overrightarrow{CD}$ + $\overrightarrow{DB}$

    c) What do you notice?

2 On a square grid with x and y from 0 to 8, plot A(2, 3) and B(3, 1).

    a) Write down $\overrightarrow{AB}$ as a column vector.

    b) Mark any two points as C and D and work out these.

      (i) $\overrightarrow{AC}$ + $\overrightarrow{CB}$

      (ii) $\overrightarrow{AD}$ + $\overrightarrow{DB}$

      (iii) $\overrightarrow{AC}$ + $\overrightarrow{CD}$ + $\overrightarrow{DB}$

    c) What do you notice?

3 The vectors **a** and **b** are drawn on the grid.
Draw the resultant of **a** and **b**.

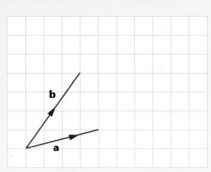

4 The vectors **a** and **b** are drawn on the grid.
Draw the resultant of 2**a** and **b**.

5 A is the point ($^-$2, 1), B is the point (4, 3) and C is the point (7, 4).

   a) Work out these column vectors.

     (i) $\overrightarrow{AB}$

     (ii) $\overrightarrow{BC}$

   b) What can you say about A, B and C?

6 A is the point (2, 1), B is the point (4, 4), C is the point (7, 4) and D is the point (3, $^-$2).

   a) Work out these column vectors.

     (i) $\overrightarrow{AB}$

     (ii) $\overrightarrow{CD}$

   b) What can you say about AB and CD?

7 In the triangle, $\overrightarrow{AB}$ = **a** and $\overrightarrow{AC}$ = 2**b**.
Find the vector $\overrightarrow{BC}$ in terms of **a** and **b**.

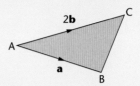

8 In the triangle ABC, $\overrightarrow{AB}$ = 2**a** and $\overrightarrow{CB}$ = 3**b**.
Work out the vector $\overrightarrow{AC}$.

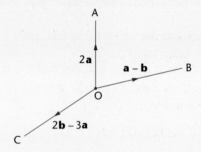

9 In this diagram, $\overrightarrow{OA}$ = 2**a**, $\overrightarrow{OB}$ = **a** – **b**, and $\overrightarrow{OC}$ = 2**b** – 3**a**.

Write these vectors in terms of **a** and/or **b**, as simply as possible:

   a) $\overrightarrow{AB}$

   b) $\overrightarrow{BC}$

   c) $\overrightarrow{AC}$

10 ABCD is a parallelogram. $\overrightarrow{AB}$ = **a** and $\overrightarrow{AD}$ = **b**.

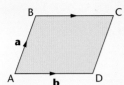

Work out the vectors $\overrightarrow{BC}$, $\overrightarrow{CD}$, $\overrightarrow{BD}$, and $\overrightarrow{AC}$ in terms of **a** and/or **b**.

11 In the triangle OAB, C is a point on AB so that AC = 2 × CB.

$\overrightarrow{OA}$ = **a** and $\overrightarrow{OB}$ = **b**.

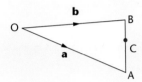

Work out the vectors $\overrightarrow{AB}$, $\overrightarrow{CB}$ and $\overrightarrow{OC}$ in terms of **a** and/or **b**.

12 ABCD is a parallelogram. E is the midpoint of the DC line.

$\overrightarrow{AB}$ = **a** and $\overrightarrow{AD}$ = **b**.

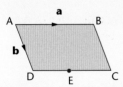

Write down the vector $\overrightarrow{EB}$ in terms of **a** and/or **b**.

13 In the trapezium ABCD, AD is parallel to BC and AD = 2 × BC.

$\overrightarrow{AB}$ = **a** and $\overrightarrow{AD}$ = **b**.

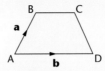

Write down the vector $\overrightarrow{CD}$ in terms of **a** and **b**.

14 Triangle AEF is a 3 times enlargement of triangle ABC.

$\overrightarrow{AB}$ = **a** and $\overrightarrow{AC}$ = **b**.

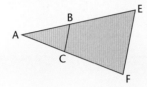

a) Write down the vectors $\overrightarrow{AE}$, $\overrightarrow{AF}$, $\overrightarrow{BC}$ and $\overrightarrow{EF}$ in terms of **a** and/or **b**.
b) What do the vectors show about BC and EF?

15 In this diagram, $\overrightarrow{OA}$ = 2**a**, $\overrightarrow{OB}$ = 2**a** + 3**b** and $\overrightarrow{OC}$ = 3**b**.

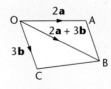

a) Write down the vectors $\overrightarrow{AB}$ and $\overrightarrow{BC}$ in terms of **a** and/or **b**.
b) What can you say about the shape OABC?

16 Work out the vector $\overrightarrow{AB}$ for this shape.

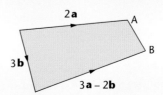

17 OAB is a triangle with E a point on OA so that OE = 2 × EA.

$\overrightarrow{OA}$ = **a** and $\overrightarrow{OB}$ = **b**.

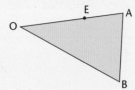

Work out the vector $\overrightarrow{EB}$ in terms of **a** and/or **b**.

18 ABCD is a kite.
E is the point where the diagonals cross.
BE = ED and CE = 3 × AE in length.

$\overrightarrow{AB}$ = **a** and $\overrightarrow{AD}$ = **b**.

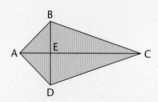

a) Work out the vectors $\overrightarrow{BD}$, $\overrightarrow{BE}$, $\overrightarrow{AE}$, $\overrightarrow{EC}$ and $\overrightarrow{BC}$ in terms of **a** and/or **b**.

b) Explain why the vectors show that BC is not parallel to AD.

19 In the triangle OCD, AC = 3 × OA and BD = 3 × OB.

$\overrightarrow{OA}$ = **a** and $\overrightarrow{OB}$ = **b**.

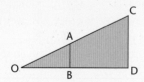

a) Use vectors to show that AB is parallel to CD.
b) What is the ratio of the lengths of AB and CD?

20 ABCDEF is a regular hexagon.
O is the centre of the hexagon.

$\overrightarrow{OA}$ = **a** and $\overrightarrow{OB}$ = **b**.

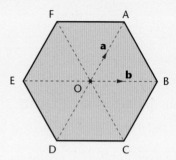

Find each of these vectors in terms of **a** and/or **b**.

a) $\overrightarrow{FA}$

b) $\overrightarrow{BD}$

c) $\overrightarrow{AB}$

d) $\overrightarrow{AC}$

21 In the diagram, P is one third of the way along AB.

$\overrightarrow{OA}$ = **a** and $\overrightarrow{OB}$ = **b**.

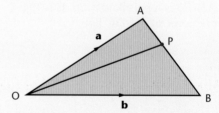

Find each of these vectors, as simply as possible, in terms of **a** and **b**.

a) $\overrightarrow{AB}$

b) $\overrightarrow{AP}$

c) $\overrightarrow{OP}$

22 In the diagram, A and B are the midpoints of OC and OD respectively.

$\overrightarrow{OA} = \mathbf{a}$ and $\overrightarrow{OB} = \mathbf{b}$.

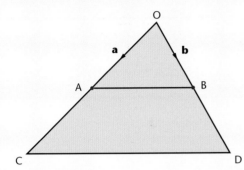

a) Find each of these vectors in terms of **a** and **b**.

   (i) $\overrightarrow{AB}$

   (ii) $\overrightarrow{CD}$

b) What can you conclude about the lines AB and CD?

23 In the diagram,

$\overrightarrow{OA} = \overrightarrow{AD} = \overrightarrow{CB} = \overrightarrow{BE} = \mathbf{a}$

and $\overrightarrow{OC} = \overrightarrow{AB} = \overrightarrow{DE} = \mathbf{c}$.

F is one third of the way along AC.

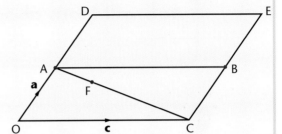

a) Find each of these vectors in terms of **a** and **c**. Simplify your answers where possible.

   (i) $\overrightarrow{OE}$

   (ii) $\overrightarrow{AC}$

   (iii) $\overrightarrow{OF}$

b) What two facts can you conclude about the points O, F and E?

K **KEY IDEAS**

■ A vector has magnitude (length) and direction but can start at any point.

■ Lines that are parallel have vectors that are multiples of each other.

■ If $\overrightarrow{BC} = n \times \overrightarrow{AB}$, then ABC is a straight line and BC is n times AB in length.

■ If $\overrightarrow{CD} = n \times \overrightarrow{AB}$, then AB and CD are parallel and CD is n times AB in length.

■ To add or subtract column vectors, add or subtract the two components separately.

■ To multiply a column vector by a scalar, multiply each component by the scalar.

■ The resultant of two vectors is the third side of the triangle formed by those vectors.

■ The vector $\overrightarrow{AB}$ is equal to the sum of the vectors $\overrightarrow{AC} + \overrightarrow{CD} + ... + \overrightarrow{PQ} + \overrightarrow{QB}$, where C, D, ..., P and Q are any points.

Comparing sets of data

Often, the purpose of calculating some statistics for a distribution is to be able to compare the distribution with others. Are the people in this group taller or shorter than average? Do people shopping at this centre on Saturdays spend more than those shopping on Tuesdays? How does the life of these light-bulbs compare with the previous design produced by this company?

The valid interpretation of the statistics you have calculated is the most important part of any statistics project. This chapter gives you more practice in these skills.

When comparing sets of data you usually need to compare two types of statistics:

■ averages
■ spread.

When comparing, you need to compare the same types of data in both distributions. For instance, comparing the mean in one with the median in the other will not tell you anything helpful.

STAGE
10

EXAM TIP
Make sure that any comparisons you make are related to the context.

A ACTIVITY 1

Work in small groups.

Choose one of these ideas.

- In your year group, are the boys taller than the girls?
- Are people's speeds of reaction determined by gender or by age?

Make notes on what data you would need to collect and how you would collect it.
What would you be looking for to make a conclusion to your chosen problem?

▌▌ EXAMPLE 1

These two box plots summarise data about the lifetimes of two different types of torch battery.

Which type of battery would you choose, and why?

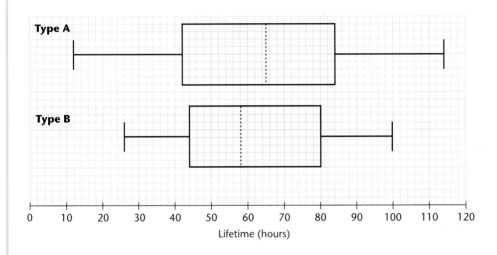

Here are some possible responses.

- I would choose type A since it has a greater median lifetime, showing it lasts longer on average.

- Type B – although its median lifetime is less, its smaller range shows that its performance is more consistent.

- Type B – I need to be able to rely on the battery lasting a good length of time, and one battery of type A lasted only 12 hours.

STAGE
10

When comparing two distributions, ideas of **skewness** may help you, although they are not needed for GCSE.

Look at the shapes of these three histograms.

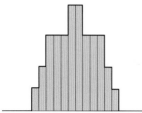 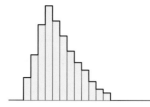 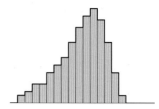

This is a symmetrical distribution.

mean = median = mode

In this distribution, the longer tail lies to the right.

This is called a positive skew.

mode < median < mean

In this distribution, the longer tail lies to the left.

This is called a negative skew.

mean < median < mode

Skewness may also be identified from the median and quartiles, for instance using box plots.

The box below shows a symmetrical distribution.

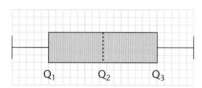

$$Q_3 - Q_2 = Q_2 - Q_1$$

The distribution below has a positive skew.

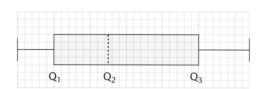

The gap in the box to the right of the median is larger than the gap to the left.

$$Q_3 - Q_2 > Q_2 - Q_1$$

The distribution illustrated below has a negative skew.

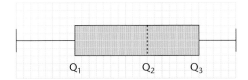

The gap in the box to the right of the median is smaller than the gap to the left.

$Q_3 - Q_2 < Q_2 - Q_1$

Sometimes, graphs in newspaper articles or in advertisements are quite complicated.

You need to be able to look for the main features of what the graphs tell you. You may sometimes find that claims are made which are not backed up by the graphs!

> **EXAM TIP**
>
> Look at the title of a graph and at the labels on the axes. Then look at the information shown by the plotted points and any trend lines.

 ACTIVITY 2

a) Look at the graph and write down the answers to these questions.

(i) Describe fully what the peak of the red graph shows.

(ii) What was the lowest value for the 'cars and light trucks' graph during the 1990s?

(iii) Criticise two aspects of the presentation of this graph.

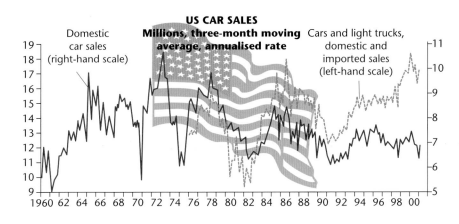

b) Working in pairs or in a group, write more questions about this graph, and answer each other's questions.

STAGE
10

A   ACTIVITY 3

Collect graphs from newspapers and magazines. Look in particular for graphs which are designed to make comparisons.

Work in groups, discussing how much information is given on the graphs. For instance, can you interpret them without needing to refer to the accompanying articles?

Say also what could be done to improve the presentation and usefulness of the graphs.

You may like to make a poster of your findings.

▌▌▌ EXERCISE 11.1

1 The masses of 400 potatoes of each of two varieties were measured.
Here are the results.

Mass (*m* g)	Frequency for variety A	Frequency for variety B
$50 < m \leqslant 100$	0	28
$100 < m \leqslant 150$	43	65
$150 < m \leqslant 200$	88	96
$200 < m \leqslant 250$	137	89
$250 < m \leqslant 300$	79	75
$300 < m \leqslant 350$	53	47

a) Draw a cumulative graph, showing both of these distributions on the same diagram.
b) Compare the distributions.

2 The masses of some tomatoes of variety A are shown in the table.

Mass (m g)	Frequency
$0 < m \leqslant 20$	3
$20 < m \leqslant 30$	6
$30 < m \leqslant 40$	11
$40 < m \leqslant 50$	18
$50 < m \leqslant 60$	10
$60 < m \leqslant 70$	2

a) Draw a cumulative frequency diagram, with a box plot below it, to represent this distribution.

Another sample of tomatoes of variety B had these summary data.

Number in sample   30
Median   46 g
Lower quartile   28 g
Upper quartile   53 g

b) Make three comparisons between these distributions.

3 These box plots show the temperatures at noon in Guildford and Torquay for one month.

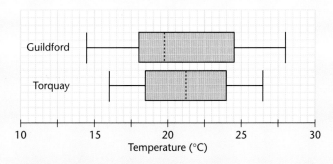

a) Which town was generally hotter?
Show how you decide.
b) Comment on the variability of temperature in the two towns.

STAGE

10

EXERCISE 11.1 continued

4   This histogram shows the population living in Aveford.

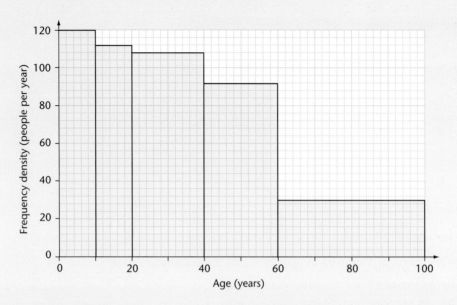

a)   How many people between the ages of 20 and 40 live in Aveford?

The table below shows the population living in Banton.

Age (a years)	$0 \leqslant a < 10$	$10 \leqslant a < 20$	$20 \leqslant a < 40$	$40 \leqslant a < 60$	$60 \leqslant a < 100$
Frequency	760	920	1680	2040	1200

b)   Construct a histogram to illustrate these data.
c)   Make two comparisons between the populations of Aveford and Banton.

5   Here are the lengths of a sample of two varieties of runner bean, Longerpod and Red Queen.

Length (L cm)	Frequency for Longerpod	Frequency for Red Queen
$10 < L \leqslant 15$	0	3
$15 < L \leqslant 20$	14	12
$20 < L \leqslant 25$	27	18
$25 < L \leqslant 30$	17	22
$30 < L \leqslant 35$	2	5

a)   Draw a cumulative graph, showing both of these distributions on the same diagram.
b)   Compare the distributions.

6 Mr Banks decided to check on the lengths of phone calls made by his son Simon and daughter Joanne. He kept a check for 40 of each of their calls. The graph shows the cumulative frequency of the lengths of each of their calls.

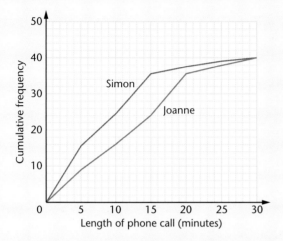

Use the median and interquartile range to compare the length of their calls.

7 The cumulative frequency graph shows the pocket money that two classes, A and B, of children have each week.

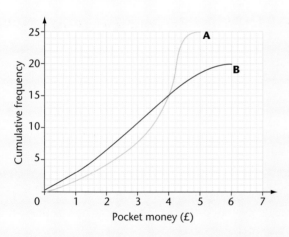

a) What information is given by the point where the two graphs cross each other?

b) Make two other comparisons between these two distributions.

8 These two histograms show the ages of passengers on a plane to the Bahamas and on a plane to Majorca.

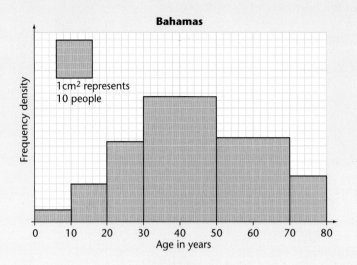

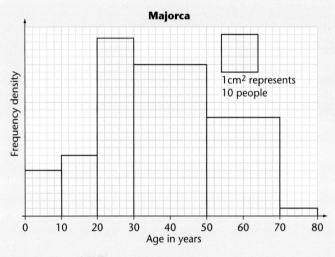

a) Comment on the relative shape of the histograms.
b) Construct frequency tables and calculate an estimate of the mean age of the passengers on each plane.

9 The graphs show the results of 700 candidates in their English and Maths examinations.

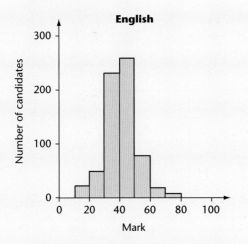

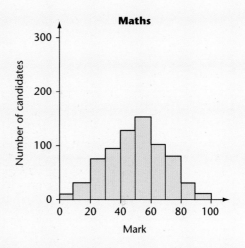

Comment on what the relative shapes of the histograms tell you.

10 To compare the length of different types of leaves, some lilac and lime leaves were measured correct to the nearest millimetre.

Length (*l* mm)	Frequency for lilac	Frequency for lime
$30 < l \leqslant 40$	0	0
$40 < l \leqslant 50$	3	2
$50 < l \leqslant 60$	3	3
$60 < l \leqslant 70$	4	4
$70 < l \leqslant 80$	7	13
$80 < l \leqslant 90$	11	12
$90 < l \leqslant 100$	12	12
$100 < l \leqslant 110$	7	4
$110 < l \leqslant 120$	3	0

Compare these distributions, making appropriate calculations and/or drawing appropriate graphs to enable you to do so.

STAGE

10

11 The heights of students at a school were measured to the nearest centimetre.
The histogram shows the heights of students in one year group.

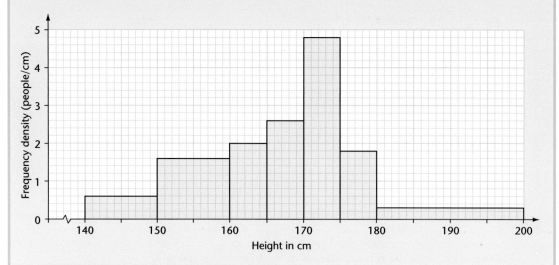

This table shows the heights of students in another year group.

Height (h cm)	Frequency
$140 < h \leq 150$	8
$150 < h \leq 155$	22
$155 < h \leq 160$	21
$160 < h \leq 165$	16
$165 < h \leq 170$	9
$170 < h \leq 175$	8
$175 < h \leq 190$	3

Compare the mean heights of the two groups.

12 These two histograms show the age distribution of the Wellfit and Superhealth fitness clubs.

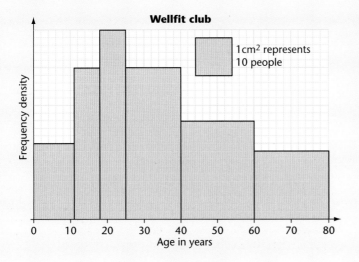

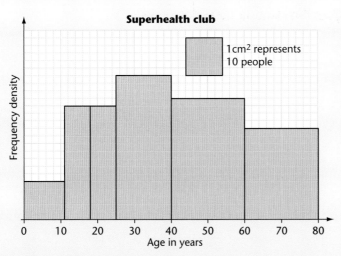

a) How many members does each of these clubs have?
b) Make two other comparisons between the memberships of the two clubs.

13 A company tested a sample of 500 light bulbs of each of two types it produces.
The table summarises the results, showing the time, in hours, that each light bulb lasted.

Time (t hours)	Frequency for type A	Frequency for type B
$0 < t \leqslant 250$	12	2
$250 < t \leqslant 500$	88	58
$500 < t \leqslant 750$	146	185
$750 < t \leqslant 1000$	184	223
$1000 < t \leqslant 1250$	63	29
$1250 < t \leqslant 1500$	7	3

a) On the same axes, draw cumulative frequency graphs to represent these distributions.
Below your graphs, draw box plots for the distributions.

b) Which of the two types of light bulb is more reliable?

14 The cumulative frequency graph shows the heights of a class of children.

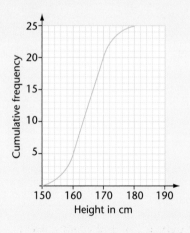

a) If, a year later, they have all grown by 4 cm, describe the shape of the cumulative frequency graph then, compared with now.

b) If, instead, the shorter children grow by 3 cm but the tallest children grow by 6 cm, draw the graphs of now and in a year's time on the same set of axes, clearly showing the comparison.

15 The histograms represent the time spent on mobile phone calls for a sample of girls and boys one week.

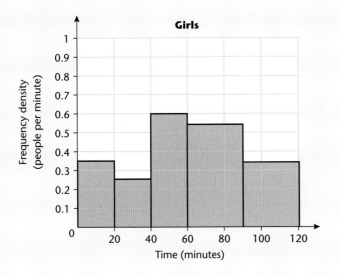

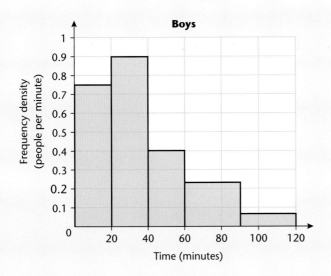

a) Find how many girls and how many boys spent between 20 and 40 minutes on the phone.

b) Compare the distributions.

STAGE

10

EXERCISE 11.1 continued

16 During the foot and mouth epidemic of 2001, a newspaper published these graphs on
9 April, comparing it with the outbreak of the disease in 1967.

How the total compares to 1967

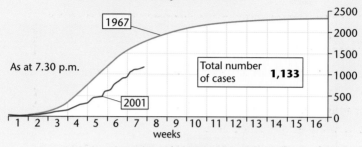

Daily outbreaks

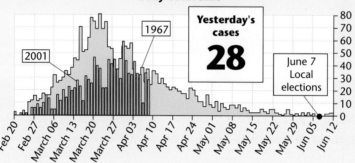

a) Describe the 1967 epidemic, as shown by these graphs.
b) Compare the 2001 outbreak with the 1967 one.

K KEY IDEAS

When comparing sets of data

■ compare averages and compare spreads.

■ compare the same type of data in each distribution.

■ use calculations and/or statistical diagrams.

■ relate your comparisons to the context of the data.

Simultaneous equations

You will learn about

- Solving two simultaneous equations where one of the equations is a quadratic equation
- Solving two simultaneous equations where one of the equations is the equation of a circle with the origin as the centre

You should already know

- How to solve linear simultaneous equations algebraically
- How to solve quadratic equations algebraically
- How to solve quadratic equations graphically
- That the equation of a circle radius r, with centre at the origin, is $x^2 + y^2 = r^2$

Solving simultaneous linear equations

 A **ACTIVITY 1**

Remind yourself about simultaneous linear equations.

Solve these using an algebraic method.

a) $x - 3y = 1$
$2x + y = 9$

b) $2x - 3y = 0$
$3x + 4y = 17$

c) $7x + 2y = 11$
$3x - 5y = {}^-7$

d) $5x + 3y = 9$
$3x - 2y = 13$

e) $4x + 3y = 5$
$6x + 7y = 10$

STAGE
10

EXAM TIP
Always check the answers to the simultaneous equations. If there are any errors, check the adding or subtracting as this is where most errors occur.

In Stage 8 you learned how to solve a pair of simultaneous linear equations by the method of elimination.

You probably used this method in Activity 1. However, Examples 1 and 2 show an alternative method, the method of substitution.

▌▌ EXAMPLE 1

Solve these simultaneous equations.

$2x + 3y = 13, 6x + 2y = 11$

Method 1: Elimination

$2x + 3y = 13$ [1]
$6x + 2y = 11$ [2]

To solve by elimination the easiest way is to multiply [1] by 3 and then subtract.

[1] × 3 $6x + 9y = 39$ [3]

                  $6x + 2y = 11$ [2]

[3] − [2] $7y = 28$

                    $y = 4$

Substitute in [1] $2x + 12 = 13$

                    $2x = 1$

                     $x = \frac{1}{2}$

Check in [2] $6x + 2y = 3 + 8 = 11$ ✓

Solution: $x = \frac{1}{2}, y = 4$

Method 2: Substitution

$2x + 3y = 13$ [1]
$6x + 2y = 11$ [2]

First rearrange [1] to make x the subject.

$$2x + 3y = 13$$
$$2x = 13 - 3y$$
$$x = \frac{13 - 3y}{2}$$

Substitute in [2]

$$6\left(\frac{13 - 3y}{2}\right) + 2y = 11$$
$$39 - 9y + 2y = 11 \qquad \text{Cancel 6 and 2 and expand the bracket.}$$
$$39 - 7y = 11$$
$$7y = 28$$
$$y = 4$$

The rest is the same as in Method 1.

Before looking at harder simultaneous equations it is best to see another example of a pair of simultaneous linear equations being solved by substitution.

▌▌ EXAMPLE 2

Solve these simultaneous equations by substitution.

$3x + 2y = 12$ [1]
$5x - 3y = 1$ [2]

Either x or y can be substituted.

Here it is easiest to substitute for y from [1] into [2].

[1] gives
$$2y = 12 - 3x$$
$$y = \frac{12 - 3x}{2}$$

Substitute in [2]

$$5x - 3\left(\frac{12 - 3x}{2}\right) = 1$$
$$10x - 3(12 - 3x) = 2 \qquad \text{Multiply through by 2.}$$
$$10x - 36 + 9x = 2$$
$$19x = 38$$
$$x = 2$$

Substitute in [1]

$$6 + 2y = 12$$
$$2y = 6$$
$$y = 3$$

Check in [2] $5x - 3y = 10 - 9 = 1$ ✓

Solution: $x = 2, y = 3$

STAGE
10

EXERCISE 12.1

Solve these simultaneous equations by the method of substitution.

1 $y = 2x - 1$
   $x + 2y = 8$

2 $3y = 11 - x$
   $3x - y = 3$

3 $3x + 2y = 7$
   $2x - 3y = {}^-4$

4 $3x - 2y = 3$
   $2x - y = 4$

5 $y = 2x - 3$
   $7x - 4y = 10$

6 $4x - 2y = 3$
   $x - y = 0$

7 $3x - y = 7$
   $5x + 2y = 8$

8 $2x + 3y = 7$
   $5y = 11 - 3x$

The intersection of a quadratic curve and a line

One harder type of simultaneous equations to solve includes one linear equation and one quadratic, for example $y = 3x + 2$ and $y = x^2 - 2x + 8$. In Chapter 1 you solved these by a graphical method. To solve these algebraically, the substitution method must be used.

EXAMPLE 3

Solve these simultaneous equations.

$y = x^2 + 3x - 7$ [1]
$y = x - 4$ [2]

a) First solve them graphically using values of x from $^-5$ to $+2$.

b) Then solve them algebraically.

a) $y = x^2 + 3x - 7$

x	-5	-4	-3	-2	-1	0	1	2
x^2	25	16	9	4	1	0	1	4
+ 3x	-15	-12	-9	-6	-3	0	3	6
- 7	-7	-7	-7	-7	-7	-7	-7	-7
$y = x^2 + 3x - 7$	3	-3	-7	-9	-9	-7	-3	3

EXAMPLE 3 continued

$y = x - 4$

x	⁻5	0	2
y	⁻9	⁻4	⁻2

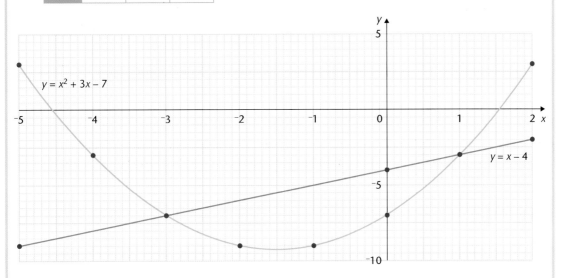

The curve and the line cross at $x = 1$, $y = ⁻3$ and at $x = ⁻3$, $y = ⁻7$.

b) Substitute for y from [2] into [1].

Substituting gives $x - 4 = x^2 + 3x - 7$

                         $x^2 + 2x - 3 = 0$ Rearrange.

                   $(x - 1)(x + 3) = 0$ Factorise.

                             $x = 1$ or $⁻3$

Substitute in [1]   For $x = 1$, $y = 1 + 3 - 7 = ⁻3$

                   For $x = ⁻3$, $y = 9 - 9 - 7 = ⁻7$

Check in [2]   For $x = 1$, $y = 1 - 4 = ⁻3$ ✓

                  For $x = ⁻3$, $y = ⁻3 - 4 = ⁻7$ ✓

Solution: $x = 1$, $y = ⁻3$ or $x = ⁻3$, $y = ⁻7$

EXAMPLE 4

Solve these simultaneous equations algebraically.

$3x + 2y = 7$ [1]
$y = x^2 - 2x + 3$ [2]

Either x or y can be substituted but the easiest way is to substitute for y from [2] into [1].

$$3x + 2(x^2 - 2x + 3) = 7$$

$$3x + 2x^2 - 4x + 6 = 7$$

$$2x^2 - x - 1 = 0$$

$$(2x + 1)(x - 1) = 0$$

$$x = \tfrac{-1}{2} \text{ or } 1$$

EXAM TIP

Always substitute for the letter that needs the least manipulation, usually y.

Substitute in [1]   For $x = \tfrac{-1}{2}$, $\tfrac{-3}{2} + 2y = 7$, $2y = 8\tfrac{1}{2}$, $y = 4\tfrac{1}{4}$

For $x = 1$, $3 + 2y = 7$, $2y = 4$, $y = 2$

Check in [2]   For $x = \tfrac{-1}{2}$, $y = \left(\tfrac{-1}{2}\right)^2 - 2\left(\tfrac{-1}{2}\right) + 3 = \tfrac{1}{4} + 1 + 3 = 4\tfrac{1}{4}$ ✓

For $x = 1$, $y = 1 - 2 + 3 = 2$ ✓

Solution: $x = \tfrac{-1}{2}, y = 4\tfrac{1}{4}$ or $x = 1, y = 2$

EXERCISE 12.2

Solve these simultaneous equations by the method of substitution.

1 $y = 10 - 2x$
$y = x^2 - 5x + 6$

2 $y - 4x - 7 = 0$
$y = x^2 - 3x - 1$

3 $3x + 2y = 7$
$y = x^2 - x + 3$

4 $x + 2y = 8$
$y = x^2 + x + 3$

5 $y = 4 - 3x$
$y = x^2 - 6x - 6$

6 $y = 2x - 3$
$y = x^2 - 4x + 5$

7 $y = x^2 + x + 3$
$2x + y = 1$

8 $y = x^2 + x - 2$
$x + 5y + 2 = 0$

9 $y = x^2 - 5x + 5$
$2x + y = 9$

10 $y = x^2 - 3x - 1$
$4x + y = 5$

11 $y = x^2 + 3$
$y = 3x + 7$

12 $y = x^2 - 5x + 3$
$7x + 2y = 11$

13 $x + y = 3$
$y = x^2 + x$

14 $y = 3x + 1$
$y = 5x^2$

15 $y = x$
$y = x^2$

The intersection of a circle and a line

So far in this chapter you have solved equations where, graphically, a straight line intersects a parabola.

In Chapter 1 you learned how to solve equations where, graphically, a straight line intersects a circle with its centre at the origin. Simultaneous equations of this form can also be solved by substitution.

‖ EXAMPLE 5

Use algebra to solve these simultaneous equations.

$x^2 + y^2 = 25$ [1]

$y = x + 1$ [2]

Reminder:
The equation of a circle centre (0, 0) and radius r is $x^2 + y^2 = r^2$.

Substitute for y from [2] into [1].

$$x^2 + (x + 1)^2 = 25$$

$$x^2 + x^2 + 2x + 1 = 25 \qquad \text{Expand the bracket.}$$

$$2x^2 + 2x - 24 = 0 \qquad \text{Collect the terms.}$$

$$x^2 + x - 12 = 0 \qquad \text{Divide by 2.}$$

$$(x + 4)(x - 3) = 0 \qquad \text{Factorise.}$$

$$x = {}^-4 \text{ or } 3$$

Substitute in [2]   For $x = {}^-4$ $y = {}^-4 + 1 = {}^-3$

                For $x = 3$ $y = 3 + 1 = 4$

Check in [1]

         For $x = {}^-4$, $y = {}^-3$, $x^2 + y^2 = 16 + 9 = 25$ ✓

         For $x = 3$, $y = 4$, $x^2 + y^2 = 9 + 16 = 25$ ✓

Solution: $x = {}^-4$, $y = 3$ or $x = 3$, $y = 4$

STAGE

10

EXERCISE 12.3

Use algebra to solve these simultaneous equations.
Where necessary, give your answers to 2 decimal places.

1 $x^2 + y^2 = 49$
   $y = 7 - x$

2 $x^2 + y^2 = 169$
   $y = x + 7$

3 $x^2 + y^2 = 25$
   $x + y = 5$

4 $x^2 + y^2 = 100$
   $y = x + 2$

5 $x^2 + y^2 = 64$
   $y = 2x + 8$

6 $x^2 + y^2 = 4$
   $y = 2 - x$

7 $x^2 + y^2 = 225$
   $y = x + 3$

8 $x^2 + y^2 = 9$
   $x + y = 3$

9 $x^2 + y^2 = 100$
   $y = 14 - x$

10 $x^2 + y^2 = 34$
   $y = x - 2$

11 $x^2 + y^2 = 25$
   $y = x + 7$

12 $x^2 + y^2 = 100$
   $y = x - 2$

13 $x^2 + y^2 = 4$
   $y = x$

14 $x^2 + y^2 = 5$
   $y = x + 2$

15 $x^2 + y^2 = 36$
   $y = 2x - 1$

16 $x^2 + y^2 = 16$
   $y = \bar{\ }x$

17 $x^2 + y^2 = 25$
   $y = 3x + 1$

18 $x^2 + y^2 = 10$
   $y = x + 3$

K KEY IDEAS

- When solving two linear simultaneous equations algebraically, the methods of elimination or substitution can be used.

- When solving, algebraically, simultaneous equations consisting of a linear equation and a quadratic or circle equation, use substitution. Use the linear equation to find one letter in terms of the other, then substitute this in the other equation.

Revision exercise C1

1 Simplify these.

a) $\dfrac{x}{2} + \dfrac{x+2}{3}$

b) $\dfrac{2x-1}{4} - \dfrac{2x+3}{5}$

c) $\dfrac{1}{x+1} + \dfrac{2}{x-2}$

d) $\dfrac{2x}{x-1} - \dfrac{x-1}{x+2}$

e) $\dfrac{3x^2 + 9x}{x^2 + 4x + 3}$

2 Solve these equations.

a) $x(x-2) - 2x(x-3) = 12 - x^2$

b) $\dfrac{2x}{3} + \dfrac{x-2}{2} = 1$

c) $x + 1 = \dfrac{16}{x+1}$

d) $\dfrac{x^2}{3} - \dfrac{x}{3} - 4 = 0$

e) $\dfrac{1}{x+1} = \dfrac{4}{3x+2}$

f) $x + 2 = \dfrac{15}{x}$

g) $\dfrac{5}{x+1} - \dfrac{2}{x-1} = \dfrac{1}{3}$

3 Given that $\mathbf{a} = \begin{pmatrix} 1 \\ 2 \end{pmatrix}$, $\mathbf{b} = \begin{pmatrix} -2 \\ 1 \end{pmatrix}$ and $\mathbf{c} = \begin{pmatrix} -1 \\ -3 \end{pmatrix}$, work out these.

a) $2\mathbf{a}$

b) $\mathbf{a} - \mathbf{b}$

c) $\mathbf{a} - \mathbf{b} + \mathbf{c}$

d) $\mathbf{a} + 2\mathbf{b}$

e) $3\mathbf{a} + 2\mathbf{c}$

f) $\frac{1}{2}\mathbf{a}$

g) $2\mathbf{a} - 3\mathbf{c}$

h) $\frac{1}{2}\mathbf{b} - \frac{1}{2}\mathbf{c}$

i) $\mathbf{a} - \frac{1}{2}\mathbf{c} - \mathbf{b}$

4 What are the coordinates of the image when

a) the point $(-2, 1)$ is translated by $\begin{pmatrix} 1 \\ 2 \end{pmatrix}$?

b) the point $(4, 3)$ is translated by $\begin{pmatrix} -4 \\ -3 \end{pmatrix}$?

c) the point $(2, -4)$ is translated by $\begin{pmatrix} 3 \\ -1 \end{pmatrix}$?

5 In the triangle ABC, $\overrightarrow{AB} = \mathbf{a}$ and $\overrightarrow{AC} = 2\mathbf{b}$.

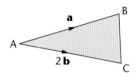

Write down the vector $\overrightarrow{BC}$ in terms of $\mathbf{a}$ and $\mathbf{b}$.

6 In the diagram, $\overrightarrow{OA} = \mathbf{a}$, $\overrightarrow{OB} = 2\mathbf{b} - \mathbf{a}$ and $\overrightarrow{OC} = 6\mathbf{b} - 5\mathbf{a}$.

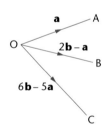

a) Work out the vectors $\overrightarrow{AB}$ and $\overrightarrow{BC}$ in terms of $\mathbf{a}$ and/or $\mathbf{b}$.

b) What can you say about AB and BC?

STAGE
10

7 ABCD is a parallelogram.
E, F, G and H are the midpoints of the sides.

$\overrightarrow{AB} = \mathbf{p}$ and $\overrightarrow{AD} = \mathbf{q}$.

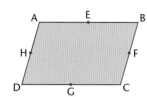

a) Find the vectors $\overrightarrow{EB}$, $\overrightarrow{BF}$, $\overrightarrow{EF}$, $\overrightarrow{HD}$, $\overrightarrow{DG}$ and $\overrightarrow{HG}$ in terms of $\mathbf{p}$ and $\mathbf{q}$.

b) What can you say about HG and EF?

8 ABCD is a rectangle.
E is a point on the diagonal AC so that AE $= 2 \times$ EC.
$\overrightarrow{AB} = \mathbf{p}$ and $\overrightarrow{AD} = \mathbf{q}$.

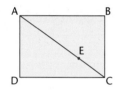

Work out the vector $\overrightarrow{EB}$.

9 ABCD is a quadrilateral with $\overrightarrow{AB} = 3\mathbf{p}$, $\overrightarrow{AD} = \mathbf{q}$ and $\overrightarrow{BC} = \mathbf{q} + 2\mathbf{p}$.
Use vectors to identify the type of quadrilateral.

10 The table shows the prices of a sample of 100 houses in the north-west of England.

Price (£000)	Number of houses
$150 < x \leqslant 170$	4
$170 < x \leqslant 190$	15
$190 < x \leqslant 210$	27
$210 < x \leqslant 230$	41
$230 < x \leqslant 250$	10
$250 < x \leqslant 270$	3

Find the median and interquartile range for this sample.
A similar sample in the south-east gave a median of £280 000 and an interquartile range of £190 000.
Compare the two areas.

11 These cumulative frequency diagrams show the marks obtained in examinations in French and English by 200 students in Year 8.

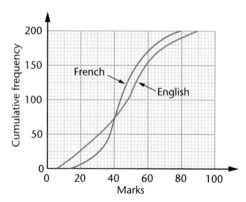

a) Draw box plots for each of the languages.

b) Use the median and interquartile range for each subject to compare the results.

12 The heights of students in two classes are measured. The results are given in the tables below.

Class 11A	
Height (H cm)	**Frequency**
$130 \leqslant H < 140$	1
$140 \leqslant H < 150$	4
$150 \leqslant H < 160$	9
$160 \leqslant H < 170$	8
$170 \leqslant H < 180$	2
$180 \leqslant H < 190$	2

Class 11B	
Height (H cm)	**Frequency**
$120 \leqslant H < 130$	4
$130 \leqslant H < 140$	5
$140 \leqslant H < 150$	8
$150 \leqslant H < 160$	3
$160 \leqslant H < 170$	3
$170 \leqslant H < 180$	1

a) Show the data on two histograms.
b) Compare the heights of the students in the two classes.

13 Solve these simultaneous equations.
a) $y = x^2 - 2x + 3$
$y = 2x$
b) $y = 2x^2 - 3x + 3$
$y = 3x - 1$
c) $y = x^2 - 4x + 5$
$y + 4x = 6$
d) $x^2 + y^2 = 36$
$y = x + 6$

14 Solve these simultaneous equations algebraically.
a) $y = x^2 - 3x$
$y = 8 - x$
b) $y = 2x^2 - 4x + 1$
$y = 3x - 2$
c) $y = x^2 - 5x + 5$
$x - 2y = 5$
d) $x^2 + y^2 = 29$
$x + 2y = 1$

15 Solve the simultaneous equations
$x^2 + y^2 = 9$
$y = x + 2$.
Give the answers correct to 1 decimal place.

STAGE
10

13 Trigonometrical functions

You will learn about

- Describing, drawing and sketching the sine, cosine and tangent graphs
- Other trigonometrical functions

You should already know

- How to use your calculator with trigonometrical functions
- Pythagoras' theorem

Trigonometrical functions of any angle

You have used sine, cosine and tangent in right-angled triangles, so the angles have been acute. You have also used sine and cosine with obtuse angles when using the sine and cosine rules.

ACTIVITY 1

Enter some non-acute angles in your calculator,

such as ⁻40°, 120°, 270°, 300°.

Find the sine of these angles. Then find the inverse sine of your answer.

What do you notice about your answers for inverse sine?
What do you notice about your answers for sine?

Repeat with some other angles, until you think you know what is happening.

Have you got enough values to sketch or draw accurately a graph of
$y = \sin \theta$? Or you could use a graphics calculator or graph-drawing
program to draw it.

Similarly, try this activity with cosine and tangent.

STAGE
10

If you enter any angle on your calculator, you will find it will give you a value for the sine, cosine or tangent of that angle.

In this diagram you can see that for an acute angle

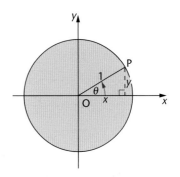

$$\cos \theta = \frac{x}{1} \quad \text{so} \quad x = \cos \theta$$

$$\sin \theta = \frac{y}{1} \quad \text{so} \quad y = \sin \theta$$

$$\tan \theta = \frac{y}{x} = \frac{\sin \theta}{\cos \theta}$$

So P has coordinates $(\cos \theta, \sin \theta)$.

For other angles, the trigonometrical functions are defined in a similar way, where the angle is measured anticlockwise from the x-axis.

In this diagram you can see that $\cos 210°$ and $\sin 210°$ are both negative.

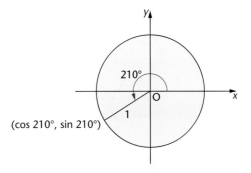

You can use symmetry to see that

$\cos 210° = {}^-\cos 30°$

$\sin 210° = {}^-\sin 30°$

$\tan 210° = \dfrac{\sin 210°}{\cos 210°} = \dfrac{\sin 30°}{\cos 30°} = \tan 30°$

So $\tan 210°$ is positive.

Continuing in a similar way, we obtain these graphs.

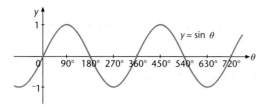

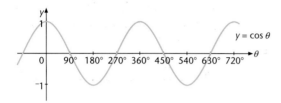

These graphs are both wave-shaped and repeat every 360°.

The length of a repeating pattern is called the **period**. Here the period is 360°.

For a wave, the amount it varies from its mean is called the **amplitude**. For these graphs the amplitude is 1.

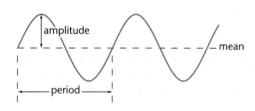

The graph of $y = \tan \theta$ has a different shape. Since $\tan \theta = \dfrac{\sin \theta}{\cos \theta}$ there is a problem when $\cos \theta = 0$.

For instance, if you try using your calculator to find $\tan 90°$, you will get an error message.
Try entering different angles from 80°, getting closer to 90°, and you will see why.

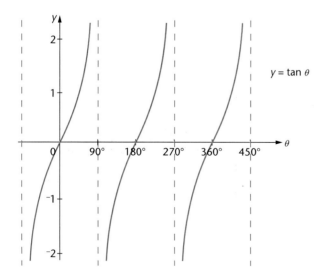

The graph of $x = \cos\theta$ shows that $\cos\theta = 0$ when $\theta = 90°, 270°, 450°$ and so on, and the graph of $y = \tan\theta$ is discontinuous at these values.

The graph of $y = \tan\theta$ is not a wave but it does repeat. Its period is $180°$. The graph approaches, but never meets, the lines $\theta = 90°$, $\theta = 270°$, and so on.

These lines are called **asymptotes** and are shown on the graph by dotted lines.

You may need to draw accurate graphs of these functions or to sketch their shapes, showing important values on the axes.

The shape of the graphs can also help you to find the values of x that satisfy equations such as $\sin x = 0.5$.

> **EXAM TIP**
>
> Learn the shapes of the graphs of
> $y = \cos x$
> $y = \sin x$
> $y = \tan x$

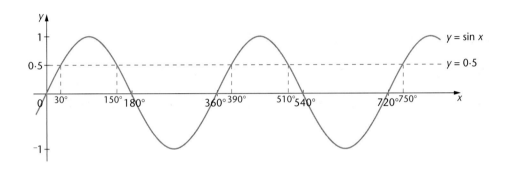

On the calculator, finding the inverse sine of 0.5 will give you the answer $30°$. However, the graph shows that there is an infinite number of solutions. Use the symmetry of the graph to see that $^{-}210°, 150°, 390°, 510°$ and so on are other solutions. You can use your calculator as a check to see that the sine of all these angles is 0.5.

EXAMPLE 1

Sketch the graph of $y = \cos x$ for values of x from $0°$ to $360°$.

Use the graph and your calculator to find two values of x between $0°$ and $360°$ for which $\cos x = {}^{-}0.8$.

Give your answers to 1 decimal place.

From the calculator, $\cos^{-1}({}^{-}0.8) = 143.1°$

$180 - 143.1 = 36.9$

From the symmetry of the graph, the other solution is

$x = 180 + 36.9 = 216.9°$

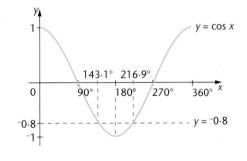

STAGE

10

EXERCISE 13.1

1 Draw accurately the graph of $y = \sin x$ for values of x from 0° to 360°, plotting values every 10°.
Use a scale of 1 cm to 20° for x and 2 cm to 1 unit for y.
Use your graph to solve these equations for $0° \leqslant x \leqslant 360°$.
a) $\sin x = 0$
b) $\sin x = 0\cdot45$
c) $\sin x = {}^-0\cdot60$

2 Draw accurately the graph of $y = \tan x$ for values of x from 0° to 360°, plotting values every 10°.
Use a scale of 1 cm to 20° for x and 2 cm to 1 unit for y, scaling y from ⁻4 to 4.
Use your graph to solve these equations for $0° \leqslant x \leqslant 360°$.
a) $\tan x = 1$
b) $\tan x = 2$

3 Sketch the graph of $y = \cos x$ for values of x from ⁻180° to 360°.
Use the graph and your calculator to find the three solutions of $\cos x = 0\cdot3$ between ⁻180° and 360°.
Give your answers to 1 decimal place.

4 Sketch the graph of $y = \tan x$ for values of x from 0° to 360°.
Use the graph and your calculator to find the two solutions of $\tan x = {}^-2$ between 0° and 360°.
Give your answers to the nearest degree.

5 On the same diagram, sketch the graphs of $y = \sin x$ and $y = \cos x$ for values of x from 0° to 360°.
State the values of x between 0° and 360° for which $\sin x = \cos x$.

6 One solution of $\cos x = 0\cdot5$ is $x = 60°$.
Without using a calculator, use the symmetry of the graph $y = \cos x$ to find the four solutions of $\cos x = {}^-0\cdot5$ between 0° and 720°.

7 Sketch the graph of $y = \sin x$ for values of x from 0 to 540°.
Use the graph and your calculator to find four solutions of $\sin x = 0\cdot8$.
Give your answers to 1 decimal place.

8 Sketch the graph of $y = \sin x$ for values of x from 0° to 360°.
Use the graph and your calculator to find the two solutions of $\sin x = {}^-0.2$ between 0° and 360°.
Give your answers to the nearest degree.

9 One value of x for which $\cos x = {}^-0\cdot3$ is 107°, to the nearest degree. Without using a calculator, use the symmetry of the graph $y = \cos x$ to find two other solutions of this equation between 0° and 540°.

10 One solution of $\tan x = 1$ is $x = 45°$. Without using a calculator, use the symmetry of the graph $y = \tan x$ to find the two solutions of $\tan x = {}^-1$ between 0° and 360°.

11 a) Draw a sketch graph of $y = \tan x$ for values of x from 0° to 360°.
b) Use your graph and a calculator to find the angles for which $\tan x = 1\cdot2$.

12 Give another three angles which have a sine value equal to each of these.
a) $\sin 20°$ **b)** $\sin 120°$
c) $\sin {}^-45°$ **d)** $\sin 390°$
e) $\sin 40°$ **f)** $\sin {}^-80°$
g) $\sin 130°$ **h)** $\sin {}^-120°$

13 Give three other angles which have a cosine value equal to each of these.
a) $\cos 140°$ **b)** $\cos {}^-120°$
c) $\cos 40°$ **d)** $\cos 90°$
e) $\cos 285°$

EXERCISE 13.1 continued

14 Give three other angles which have a tangent value equal to each of these.
  a) $\tan 45°$
  b) $\tan 120°$
  c) $\tan 40°$
  d) $\tan {}^{-}80°$
  e) $\tan 30°$
  f) $\tan 135°$

15 For $0° \leqslant x \leqslant 360°$, solve each of these.
  a) $\sin x = {}^{-}0\cdot37$
  b) $\cos x + 2 = 3$
  c) $\sin x = \tan x$
  d) $2\sin x = 1$
  e) $\tan x = {}^{-}1\cdot56$
  f) $\sin x + 2 = 1\cdot5$
  g) $\cos x = \sin x$
  h) $3\cos x = 2$

Other trigonometrical graphs

ACTIVITY 2

Use a graphics calculator or a graph-drawing computer program for this activity.

a) Draw the graph of $y = \sin x$ for values of x from $^{-}180°$ to $540°$.

On the same axes, rescaling as necessary, draw these graphs.

$y = 2\sin x \quad y = 3\sin x \quad y = {}^{-}2\sin x \quad y = 0\cdot5\sin x$

Experiment further till you think you know what the graph of $y = a\sin x$ looks like for any value of a.

Can you say how the graph of $y = \sin x$ has been transformed?

b) Clear the screen and try the graphs of $y = \cos x$, $y = 2\cos x$ and so on.

Can you describe what happens? What about $y = \tan x$, $y = 2\tan x$ and so on?

c) Clear the screen and draw again the graph of $y = \sin x$ for values of x from $^{-}180°$ to $540°$.

On the same axes, draw the graph of $y = \sin 2x$. This notation means $y = \sin(2x)$, but you will probably not need the brackets for your program. Draw also the graphs of $y = \sin 3x$, $y = \sin 0.5x$ and $y = \sin(^{-}x)$.

Experiment further till you think you know what the graph of $y = \sin ax$ looks like for any value of a.

d) What about the graphs of $y = \cos ax$ or $y = \tan ax$?

Can you describe how the graphs of $y = \cos x$ or $y = \tan x$ are transformed to these graphs?

If, instead of plotting $y = \sin x$, you plot
$y = 3 \sin x$, what difference does it make?

Here are both of these graphs plotted
together.

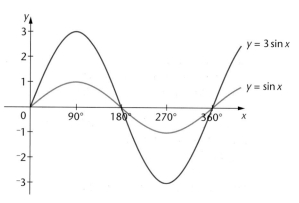

The period has stayed the same, $360°$,
but the amplitude has increased from 1 to 3.

Similarly the graph of $y = 5 \cos x$ has a period
of $360°$ but an amplitude of 5.

The graph of $y = \cos 3x$ is different. The notation $\cos 3x$ means $\cos(3x)$.

Here are the graphs of $y = \cos x$ and
$y = \cos 3x$ plotted together for
comparison.

$y = \cos 3x$ has an amplitude of 1 but
a period of $\dfrac{360°}{3} = 120°$.

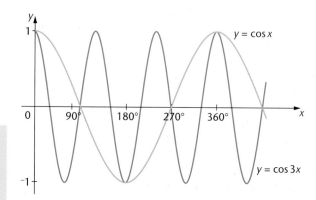

The graphs of $y = a \sin bx$ and
$y = a \cos bx$ both have amplitude
a and period $\dfrac{360°}{b}$.

EXAMPLE 2

Sketch the graph of $y = 3 \sin 2x$ for $x = 0°$ to $360°$.

Find the solutions of $3 \sin 2x = 2$ between $0°$ and $90°$, giving your answers to
1 decimal place.

$3 \sin 2x = 2$

so $\quad \sin 2x = \dfrac{2}{3}$

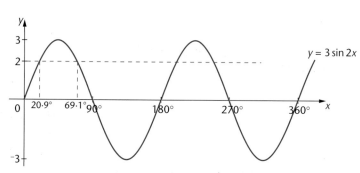

From the inverse sine function on the calculator,
$2x = 41 \cdot 810° \ldots$
$x \ = 20 \cdot 9°$ to 1 decimal place

From the symmetry of the graph, the other solution is $90° - 20 \cdot 9° = 69 \cdot 1°$.

EXERCISE 13.2

1 Draw accurately the graph of $y = 2\sin x$ for values of x from $0°$ to $180°$, plotting values every $10°$.

2 Draw accurately the graph of $y = \cos 2x$ for values of x from $0°$ to $180°$, plotting values every $10°$.

3 Draw accurately the graph of $y = 3\cos x$ for values of x from $0°$ to $180°$, plotting values every $10°$.

4 Draw accurately the graph of $y = \sin 4x$ for values of x from $0°$ to $180°$, plotting values every $10°$.

5 State the amplitude and period of each of these.

   a) $y = 3\sin x$ **b)** $y = 4\cos 2x$
   c) $y = 2\cos 0\cdot5x$ **d)** $y = 5\cos x$
   e) $y = 2\sin 3x$ **f)** $y = 4\sin \frac{1}{3}x$.

6 This is part of the graph of $y = a\cos bx$. State the values of a and b.

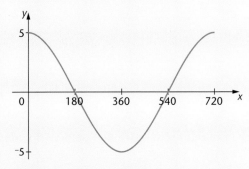

7 This is part of the graph of $y = a\sin bx$. State the values of a and b.

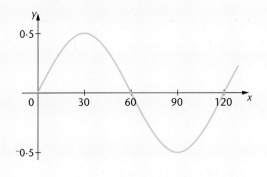

8 This is part of the graph of $y = a\cos bx$. State the values of a and b.

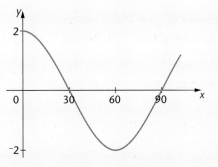

9 This is part of the graph of $y = a\sin bx$. State the values of a and b.

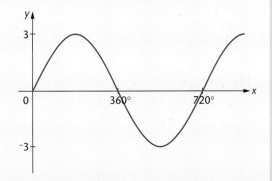

10 Sketch the graph of $y = \sin\frac{1}{2}x$ for values of x from $0°$ to $360°$.

11 Find the four solutions of $\sin 2x = 0\cdot5$ between $0°$ and $360°$.

12 Find the solutions of $\cos 2x = 0$ for $0° < x \le 360°$.

13 On the same diagram, sketch the curves of $y = \cos 2x$ and $y = \sin x$ for $0° \le x \le 90°$.
   How many solutions of the equation $\cos 2x = \sin x$ are there for $0° \le x \le 90°$?

14 Sketch the graph of $y = \cos\frac{1}{3}x$ for values of x from $0°$ to $540°$.

15 Find all the solutions of $\cos 3x = ^-1$ between $0°$ and $360°$.

16 Find the solutions of $2\sin x = 1$ for $0° < x \leqslant 360°$.

17 On the same diagram, sketch the curves $y = \sin 3x$ and $y = \cos x$ for $0° \leqslant x \leqslant 90°$. How many solutions of the equation $\sin 3x = \cos x$ are there for $0° \leqslant x \leqslant 90°$?

K KEY IDEAS

- The shapes and main features of trigonometrical graphs can be seen in these graphs.

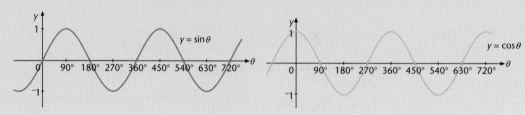

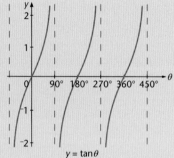

- You can find solutions to trigonometrical equations using your calculator and the symmetry of these graphs.

- The 'length' of a repeating pattern is called the period. For a wave, the amount it varies from its mean is called the amplitude.

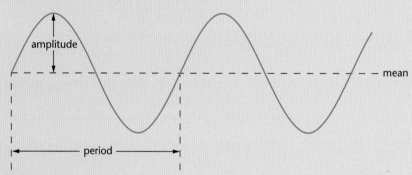

- The graphs of $y = a\sin bx$ and $y = a\cos bx$ both have amplitude a and period $\dfrac{360°}{b}$.

Transforming functions

14

14

You will learn about

- Transforming linear, quadratic, sine and cosine graphs

You should already know

- About reflections and translations
- How to find the equation of a straight line
- The shapes of graphs such as $y = x^2$, $y = x^3$, $y = \sin x$

Function notation

$y = f(x)$ means that y is a **function** of x. You read $f(x)$ as 'f of x'.

If $y = 4x - 3$ then $f(x) = 4x - 3$.

$f(2)$ means the value of the function when $x = 2$.
In this case, $f(2) = 4 \times 2 - 3 = 5$.

Function notation is a useful shorthand when several different functions are being described. As well as $f(x)$, $g(x)$ and $h(x)$ are commonly used to describe functions.

EXAMPLE 1

If $f(x) = 3x^2 - 5$, find

a) $f(2)$.

b) $f(-1)$.

a) $f(2) = 3 \times 2^2 - 5 = 7$

b) $f(-1) = 3 \times (-1)^2 - 5 = -2$

14

STAGE
10

EXAMPLE 2

If $g(x) = 5x + 6$

a) solve $g(x) = 8$.

b) write an expression for

   (i) $g(3x)$ **(ii)** $3g(x)$.

a) $g(x) = 8$
    $5x + 6 = 8$
       $5x = 2$
        $x = 0{\cdot}4$

b) (i) $g(3x) = 5(3x) + 6$
           $= 15x + 6$

    (ii) $3g(x) = 3(5x + 6)$
           $= 15x + 18$

Translations

ACTIVITY 1

If possible, use graph-drawing software to draw the graphs in this task and print them out. Otherwise, draw the graphs on graph paper.

If your grid gets too crowded, start a new one.

Section 1

a) For the function $f(x) = x^2$
- plot the graph of $y = f(x)$.
- on the same axes, plot the graph of $y = f(x) + 2$.

b) What transformation maps $y = f(x)$ on to $y = f(x) + 2$?

c) On the same axes, plot some other graphs of the form $y = f(x) + a$, where a can take any value. Try values of a that are positive, negative and fractional.

d) Describe the transformation that maps the graph of $y = f(x)$ on to $y = f(x) + a$.

A ACTIVITY 1 continued

Section 2

a) For the function $f(x) = x^2$
  - plot the graph of $y = f(x)$ on a new grid.
  - on the same axes, plot the graph of $y = f(x + 1)$.

b) What transformation maps $y = f(x)$ on to $y = f(x + 1)$?

c) On the same axes, plot $y = f(x - 2)$ and $y = f(x + 2)$.

d) Experiment further until you can describe the transformation that maps the graph of $y = f(x)$ on to $y = f(x + a)$, for any value of a.

Section 3

What happens if you change the function?

Work through sections 1 and 2 again, using $y = \sin x$ for values of x from 0° to 360°.

Describe what you find.

You already know that the graph of $y = 4x$ is a straight line through the origin with gradient 4. The graph of $y = 4x + 3$ is a straight line with gradient 4 through the point $(0, 3)$.

You can think of this as a transformation by saying that the first graph has been translated by $\binom{0}{3}$ to get the second.

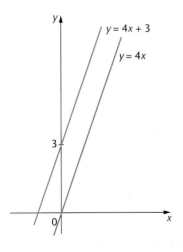

STAGE
10

Similarly, the graph of $y = x^2 - 5$ is the same shape as the graph of $y = x^2$, translated by $\begin{pmatrix} 0 \\ -5 \end{pmatrix}$.

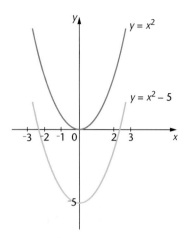

The same thing is true for all families of graphs.

The graph of $y = f(x) + a$ is the graph of $y = f(x)$ translated by $\begin{pmatrix} 0 \\ a \end{pmatrix}$.

Translating parallel to the x-axis, a different pattern emerges.

The equation $(x - 2)^2 = 0$ has root $x = 2$. Looking at the graphs of $y = x^2$ and $y = (x - 2)^2$, it is clear that $y = x^2$ has been translated by $\begin{pmatrix} 2 \\ 0 \end{pmatrix}$.

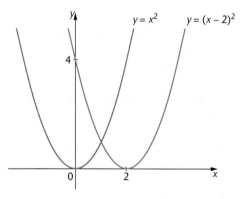

In general:

The graph of $y = f(x - a)$ is the graph of $y = f(x)$ translated by $\begin{pmatrix} a \\ 0 \end{pmatrix}$.

EXAMPLE 3

State the equation of the sine curve drawn here.

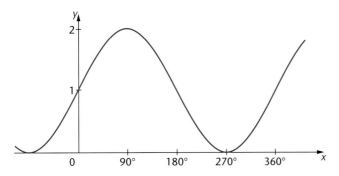

The curve of $y = \sin x$ has been translated by $\begin{pmatrix} 0 \\ 1 \end{pmatrix}$.

So its equation is $y = \sin x + 1$.

EXERCISE 14.1

1 a) Sketch these graphs on the same diagram.
  (i) $y = x^2$
  (ii) $y = x^2 + 3$
 b) State the transformation which maps **(i)** on to **(ii)**.

2 a) Sketch these graphs on the same diagram.
  (i) $y = x^2$
  (ii) $y = (x + 2)^2$
 b) What transformation maps **(i)** on to **(ii)**?

3 a) Sketch these graphs on the same diagram.
  (i) $y = x^2$
  (ii) $y = (x - 2)^2$
  (iii) $y = (x - 2)^2 + 3$
 b) What transformation maps **(i)** on to **(iii)**?

4 a) Sketch these graphs on the same diagram.
  (i) $y = {}^{-}x^2$
  (ii) $y = 2 - x^2$
 b) State the transformation which maps **(i)** on to **(ii)**.

5 a) Sketch these graphs on the same diagram.
  (i) $y = {}^{-}x^2$
  (ii) $y = {}^{-}(x + 3)^2$
 b) What transformation maps **(i)** on to **(ii)**?

6 a) Sketch these graphs on the same diagram.
  (i) $y = x^2$
  (ii) $y = (x + 2)^2$
  (iii) $y = (x + 2)^2 - 3$
 b) What transformation maps **(i)** on to **(iii)**?

STAGE
10

EXERCISE 14.1 continued

7 a) Sketch the result of translating the graph of $y = \sin x$ by $\begin{pmatrix} 0 \\ -1 \end{pmatrix}$.

b) State the equation of the transformed graph.

8 a) Sketch the result of translating the graph of $y = \cos x$ by $\begin{pmatrix} 0 \\ 1 \end{pmatrix}$.

b) State the equation of the transformed graph.

9 State the equation of the graph of $y = \sin x$ after it has been translated by

a) $\begin{pmatrix} 3 \\ 0 \end{pmatrix}$ 　　　　**b)** $\begin{pmatrix} 0 \\ 4 \end{pmatrix}$

10 State the equation of the graph of $y = x^2$ after it has been translated by

a) $\begin{pmatrix} 0 \\ -5 \end{pmatrix}$ 　　　　**b)** $\begin{pmatrix} -2 \\ 0 \end{pmatrix}$

c) $\begin{pmatrix} 1 \\ 2 \end{pmatrix}$ 　　　　**d)** $\begin{pmatrix} 3 \\ -4 \end{pmatrix}$

11 This is the graph of $y = f(x)$.

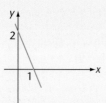

Sketch these graphs.
a) $y = f(x) - 2$
b) $y = f(x - 2)$

12 This is the graph of $y = g(x)$.

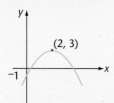

State the coordinates of the highest point on these graphs.
a) $y = g(x + 1)$
b) $y = g(x) + 1$

13 This is the graph of $y = f(x)$.

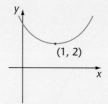

State the coordinates of the lowest point on these graphs.
a) $y = f(x) - 2$
b) $y = f(x + 2)$

14 This graph is a transformed cosine curve. State its equation.

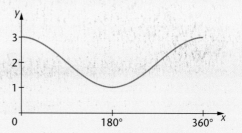

15 The graph of $y = x^2$ is translated by $\begin{pmatrix} -2 \\ 3 \end{pmatrix}$.

a) State the equation of the transformed graph.

b) Show that this equation may be written as $y = x^2 + 4x + 7$.

16 This graph is a transformed sine curve.

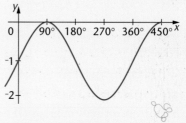

State its equation.

EXERCISE 14.1 continued

17 This is the graph of $y = g(x)$.

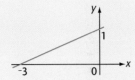

Sketch these graphs.
a) $y = g(x) - 3$
b) $y = g(x - 3)$

18 **a)** Show that the equation
$y = x^2 - 6x + 1$ may be written as
$y = (x - 3)^2 - 8$.

b) Hence state the coordinates of the
minimum point on the graph of
$y = x^2 - 6x + 1$.

One-way stretches

ACTIVITY 2

If possible, use graph-drawing software to draw the graphs in this task and print them out.
Otherwise, draw the graphs on graph paper.

If your grid gets too crowded, start a new one.

Section 1

a) For the function $f(x) = x^2 - 2x$
- plot the graph of $y = f(x)$.
- on the same axes, plot the graphs of $y = 2f(x)$ and $y = 3f(x)$.

b) Experiment further until you can describe the transformation that maps the graph of
$y = f(x)$ on to $y = kf(x)$, for any value of k.

Section 2

a) For the function $f(x) = x^2 - 2x$
- plot the graph of $y = f(x)$ on a new grid.
- on the same axes, plot the graph of $y = f(2x)$ (i.e. $y = 4x^2 - 4x$).
- also on the same axes, plot the graphs of $y = f\left(\dfrac{x}{2}\right)$ and $y = f(-3x)$.

b) Experiment further until you can describe the transformation that maps the graph of
$y = f(x)$ on to $y = f(kx)$, for any value of k.

STAGE
10

In Chapter 13 you learned about the shapes of sine and cosine graphs.

This diagram shows the graphs of $y = \sin x$ and $y = 3\sin x$.

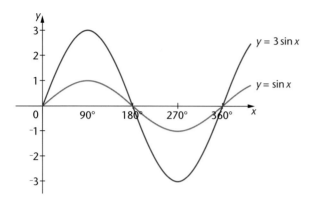

To get from $y = \sin x$ to $y = 3\sin x$, the graph has been stretched parallel to the y-axis with scale factor 3. This is an example of a general principle.

> **The graph of $y = k\text{f}(x)$ is a one-way stretch of the graph of $y = \text{f}(x)$ parallel to the y-axis with scale factor k.**

The next diagram shows the graphs of $y = \cos x$ and $y = \cos 2x$.

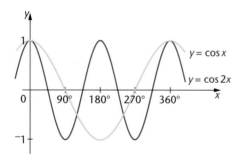

For the graph of $y = \cos 2x$ compared with the graph of $y = \cos x$, twice as much curve has been squashed into each part of the x-axis. This is described formally as a one-way stretch parallel to the x-axis with scale factor $\frac{1}{2}$. This is an example of another general principle.

> **The graph of $y = \text{f}(kx)$ is a one-way stretch of the graph of $y = \text{f}(x)$ parallel to the x-axis with scale factor $\frac{1}{k}$.**

When $k = {}^{-}1$ in these one-way stretches, there is a much simpler way of describing the transformation – as reflections. For example, the graph of $y = {}^{-}x^2$ is a reflection of $y = x^2$ in the x-axis.

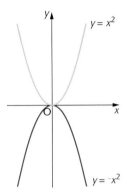

The graph of $y = {}^{-}f(x)$ is a reflection in the x-axis of the graph of $y = f(x)$.

Comparing the graphs of $y = x^3 + 1$ and $y = {}^{-}x^3 + 1$ [or $y = ({}^{-}x)^3 + 1$], you can see that one is a reflection of the other in the y-axis.

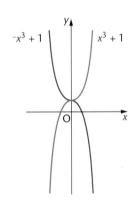

EXAM TIP
Take care not to confuse $f({}^{-}x)$ with ${}^{-}f(x)$.

The graph of $y = f({}^{-}x)$ is a reflection in the y-axis of the graph of $y = f(x)$.

EXAMPLE 4

If $f(x) = x^2 + 2$, state these equations in the form $y = ax^2 + bx + c$.

a) $y = f(4x)$

b) $y = 3f(x)$

a) $y = (4x)^2 + 2$ so $y = 16x^2 + 2$

b) $y = 3(x^2 + 2)$ so $y = 3x^2 + 6$

EXERCISE 14.2

1 a) Sketch on the same diagram the graphs of $y = \cos x$ and $y = 2 \cos x$ for $0° \leqslant x \leqslant 360°$.

b) Describe the transformation that maps $y = \cos x$ on to $y = 2 \cos x$.

2 a) Sketch on the same diagram the graphs of $y = \sin x$ and $y = \sin\frac{1}{2}x$ for $0° \leqslant x \leqslant 360°$.

b) Describe the transformation that maps $y = \sin x$ on to $y = \sin\frac{1}{2}x$.

3 Describe the transformation that maps $y = \sin x$ on to $y = \sin 3x$.

4 Describe the transformation that maps
a) $y = \sin x + 1$ on to $y = \sin(^-x) + 1$.
b) $y = x^2 + 2$ on to $y = ^-x^2 - 2$.
c) $y = x^2$ on to $y = 3x^2$.

5 a) Sketch on the same diagram the graphs of $y = \cos x$ and $y = ^-\cos x$ for $0° \leqslant x \leqslant 360°$.

b) Describe the transformation that maps $y = \cos x$ on to $y = ^-\cos x$.

6 a) Sketch on the same diagram the graphs of $y = \sin x$ and $y = \sin 2x$ for $0° \leqslant x \leqslant 360°$.

b) Describe the transformation that maps $y = \sin x$ on to $y = \sin 2x$.

7 Describe the transformation that maps $y = \sin x$ on to $y = \sin\frac{1}{3}x$.

8 Describe the transformation that maps
a) $y = \cos x + 1$ on to $y = ^-\cos x - 1$.
b) $y = x + 2$ on to $y = ^-x + 2$.
c) $y = x^2$ on to $y = 5x^2$.

9 The graph of $y = \cos x$ is transformed by a one-way stretch parallel to the x-axis with scale factor $\frac{1}{3}$.
State the equation of the resulting graph.

10 State the equation of the graph of $y = x^2 + 5$ after
a) reflection in the y-axis.
b) reflection in the x-axis.

11 State the equation of the graph of $y = x + 2$ after
a) a one-way stretch parallel to the y-axis with scale factor 3.
b) a one-way stretch parallel to the x-axis with scale factor $\frac{1}{2}$.

12 Describe transformations to map $y = g(x)$ on to each of these.
a) $y = g(x) + 1$
b) $y = 3g(x)$
c) $y = g(2x)$
d) $y = 5g(3x)$

13 State the equation of the graph of $y = 2x + 1$ after
a) reflection in the x-axis.
b) reflection in the y-axis.
c) a one-way stretch parallel to the x-axis with scale factor 0·5.

14 The graph of $y = x^2$ is stretched parallel to the x-axis with scale factor 2.
a) State the equation of the resulting graph.
b) What point does (1, 1) map on to under this transformation?
c) (i) What is the scale factor of the stretch parallel to the y-axis which maps $y = x^2$ on to the same graph?
(ii) What point does (1, 1) map on to under this transformation?

15 The graph of $y = \sin x$ is transformed by a one-way stretch parallel to the x-axis with scale factor $\frac{1}{4}$.
State the equation of the resulting graph.

EXERCISE 14.2 continued

16 State the equation of the graph of
$y = x^2 - 1$ after
  a) reflection in the y-axis.
  b) reflection in the x-axis.

17 State the equation of the graph of
$y = 4x + 1$ after
  a) a one-way stretch parallel to the
  y-axis with scale factor 4.
  b) a one-way stretch parallel to the
  x-axis with scale factor 0·5.

18 Describe transformations to map
$y = h(x)$ on to each of these.
  a) $y = h(x) - 2$
  b) $y = 3h(x)$
  c) $y = h(0·5x)$
  d) $y = 4h(2x)$

19 State the equation of the graph of
$y = x^2 + 3$ after
  a) reflection in the x-axis.
  b) reflection in the y-axis.
  c) a one-way stretch parallel to the
  x-axis with scale factor 0·5.

20 Find the equation of the graph of
$y = {}^-x^2 + 2x$ after
  a) a reflection in the x-axis.
  b) a reflection in the y-axis.
  c) a translation of 3 parallel to the
  x-axis.

21 Use the shape of the graph of $y = \sin x$
to sketch on the same axes the graphs
for $0° \leqslant x \leqslant 180°$ of
  a) $y = \sin \frac{1}{2}x$.
  b) $y = 3 \sin \frac{1}{2}x$.

22 The equation of this graph is
$y = \sin ax$. Find a.

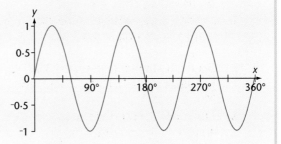

23 The equation of this graph is
$y = \cos bx$. Find b.

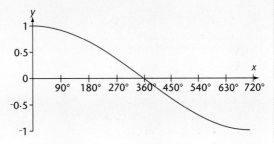

24 The equation of this graph is
$y = a \sin bx$. Find a and b.

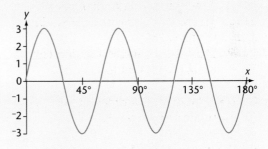

25 The equation of this graph is $y = ax^2$.
Find a.

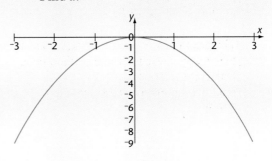

K KEY IDEAS

- The graph of $y = f(x) + a$ is the graph of $y = f(x)$ translated by $\begin{pmatrix} 0 \\ a \end{pmatrix}$.

- The graph of $y = f(x - a)$ is the graph of $y = f(x)$ translated by $\begin{pmatrix} a \\ 0 \end{pmatrix}$.

- The graph of $y = kf(x)$ is a one-way stretch of the graph of $y = f(x)$ parallel to the y-axis with scale factor k.

- The graph of $y = f(kx)$ is a one-way stretch of the graph of $y = f(x)$ parallel to the x-axis with scale factor $\dfrac{1}{k}$.

- The graph of $y = {}^-f(x)$ is a reflection in the x-axis of the graph of $y = f(x)$.

- The graph of $y = f({}^-x)$ is a reflection in the y-axis of the graph of $y = f(x)$.

STAGE
10

Probability

15

You will learn about

- Solving problems involving the addition or multiplication of two probabilities

You should already know

- The addition rule for mutually exclusive events, P(A or B) = P(A) + P(B)
- The multiplication rule for independent events, P(A and B) = P(A) × P(B)
- How to draw and use probability tree diagrams
- How to find probabilities of dependent events

In previous work on probability you learned how to calculate probabilities and how to deal with mutually exclusive events and with independent and dependent events.

- You can use the fact that the sum of the probabilities of all the mutually exclusive outcomes of an event is 1, including the fact that the probability of an event happening is 1 – the probability of it not happening.

- You use the addition rule to find the probability of either of two mutually exclusive events occurring.

P(A or B) = P(A) + P(B)

- You use the multiplication rule to find the probability of both of two events occurring.

P(A and B) = P(A) × P(B)

- You can use the multiplication rule for independent events and for dependent events, but you need to be careful about the values you use for the probabilities of dependent events. You may be able to calculate the probabilities or you may be told these.

STAGE
10

The emphasis in this chapter is on deciding which of the facts you know you need to apply in order to solve problems. You may like to use tree diagrams to help clarify the problem, but this is not essential.

EXAMPLE 1

On my journey to work on Mondays and Tuesdays I have to cross a railway line.

The probability that I get stopped at the level crossing any morning is 0·6.

What is the probability that I get stopped

a) on both days?

b) on just one of the two days?

The probability that I am stopped on Tuesday is not influenced by what happened on Monday so the events are independent.

a) Use the multiplication rule.

Probability of being stopped on both days = 0·6 × 0·6 = 0·36

b) There are two ways you can be stopped on just one day.

You need to use both the multiplication rule and the addition rule.

Probability of being stopped on just one of the days
   = P(stopped on Monday but not on Tuesday)
      + P(not stopped on Monday but stopped on Tuesday)
   = (0·6 × 0·4) + (0·4 × 0·6)
   = 0·24 + 0·24
   = 0·48

▎▎ EXAMPLE 2

Joanne is eating a box of chocolates.

She has ten chocolates left. Six are milk and four are plain.

She picks out a chocolate and eats it and then eats another.

Find the probability that

a) the first chocolate was a milk one.

b) the second chocolate was also a milk one.

c) at least one of the chocolates was plain.

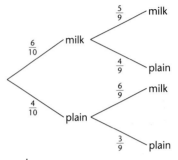

As Joanne is eating the first chocolate she chooses, she is affecting the probability of the outcome of her second choice. The events are therefore dependent events. You must work out the probabilities.

You can draw a probability tree diagram to show the possible outcomes.

a) The first part uses this fact.

The probability of an event = $\dfrac{\text{number of ways the event can happen}}{\text{total number of possible outcomes}}$

P(milk chocolate) = $\frac{6}{10}$ or 0·6

b) This asks about the probability of picking of milk chocolate given that Joanne has already eaten a milk chocolate.

There are nine chocolates left and five of them are milk chocolates.

P(second is also milk) = $\frac{5}{9}$

c) To work out the probability that at least one of the chocolates is plain you can use the multiplication rule and the addition rule.

P(at least one plain)

= P(milk followed by plain) + P(plain followed by milk) + P(plain followed by plain)

= $\frac{6}{10} \times \frac{4}{9} + \frac{4}{10} \times \frac{6}{9} + \frac{4}{10} \times \frac{3}{9}$

= $\frac{24}{90} + \frac{24}{90} + \frac{12}{90}$

= $\frac{60}{90}$

= $\frac{2}{3}$

Alternatively, you can use the fact that the probability of an event happening is 1 – the probability of it not happening.

P(at least one plain) = 1 – P(both milk)

= $1 - \frac{6}{10} \times \frac{5}{9}$

= $1 - \frac{30}{90}$

= $1 - \frac{1}{3}$

= $\frac{2}{3}$

STAGE

10

EXAMPLE 3

Dave rides his mountain bike across the moors.

When it rains the probability of him falling off is $\frac{1}{10}$.

When it is dry the probability of him falling off is $\frac{1}{50}$.

The probability that it rains when Dave goes mountain biking is $\frac{1}{5}$.

a) Find the probability that Dave goes mountain biking on a rainy day but does not fall off.

b) Find the probability that Dave falls off when he goes mountain biking.

Whether Dave falls off or not is dependent on whether or not it is raining.
You are told the probabilities.

In this example, a tree diagram is not used, but you should draw one if you find it helpful.

a) Use the multiplication rule.

P(it rains and Dave does not fall off) $= \frac{1}{5} \times \frac{9}{10}$

$$= \frac{9}{50} \text{ or } 0.18$$

b) Use the multiplication rule and the addition rule.

P(Dave falls off) = P(it rains and he falls off) + P(it doesn't rain and he falls off)

$$= \frac{1}{5} \times \frac{1}{10} + \frac{4}{5} \times \frac{1}{50}$$

$$= \frac{1}{50} + \frac{4}{250}$$

$$= \frac{5}{250} + \frac{4}{250}$$

$$= \frac{9}{250} \text{ or } 0.036$$

EXERCISE 15.1

1 A box of chocolates contains 12 hard centres and 10 soft centres.
A chocolate is taken out and eaten and then a second chocolate is taken out.
Find the probability that
   a) both chocolates have soft centres.
   b) one chocolate has a hard centre and one has a soft centre.

2 In winter the probability that it will rain on any day is $\frac{5}{7}$.
Find the probability that
   a) it will rain on two consecutive days.
   b) it will rain on the first day but not on the second.
   c) it will rain on at least one of the two days.

3 A jar contains seven red discs and four white discs.
A disc is taken from the jar and not replaced.
A second disc is now taken.
Find the probability that
  a) both discs are red.
  b) at least one disc is white.
  c) at least one disc is red.

4 The probability that a train arrives on time is 0·65.
Find the probability that for three consecutive days
  a) the train arrives on time.
  b) the train is late on the first two days.
  c) the train is on time for two out of the three days.
  d) the train is on time at least twice during the three days.

5 Mark has seven clean shirts.
Four have a striped pattern and three are plain.
  a) On Monday Mark picks a shirt at random.
    Find the probability that the shirt is plain.
  b) On Tuesday he picks another clean shirt.
    Find the probability that this shirt is striped.
  c) On Wednesday he picks a third clean shirt.
    What is the probability that this shirt is the same sort as the shirt he wore on Monday?

6 Light bulbs are packed in boxes of three.
10% of the bulbs from the production line are found to be faulty.
Calculate the probability of finding exactly two faulty bulbs in any one box.

7 Ruth has three raffle tickets.
100 tickets were sold.
  a) A ticket is drawn at random for the first prize.
    What is the probability that Ruth wins the first prize?
  b) Ruth did not win the first prize and the second ticket is drawn.
    What is the probability that Ruth wins the second prize?
  c) Ruth did win the second prize and a third ticket is drawn.
    What is the probability that Ruth wins the third prize?
  d) What is the probability that Ruth wins all three prizes?

8 Three cards are picked at random from a normal pack of playing cards, without replacement.
Find the probability of picking
  a) two clubs first.
  b) three clubs.
  c) no clubs.
  d) at least one club.

9 A box contains five red counters and seven green counters.
Will picks out a counter at random and places it on the table.
He then picks out a second counter.
  a) What is the probability that the first counter is green?
  b) The first counter was green.
    What is the probability that the second counter is also green?
  c) What is the probability that Will chooses two red counters?

10 Penny is buying two goldfish from a pet shop.
The tank of goldfish contains seven male fish and eight female fish, although they all look the same to Penny.
Find the probability that she chooses
  a) two female fish.
  b) two fish of the same sex.

STAGE 10

EXERCISE 15.1 continued

11 The pieces of a child's sorting game are shown in the table.

	Triangles	Squares	Total
Red	4	6	10
Blue	3	5	8
Total	7	11	18

However, two shapes are lost.
Find the probability that
a) they are the same shape.
b) they are the same colour.
c) they are the same shape and the same colour.

12 Helen and Katharine are going to a party.
The probability that Helen will wear a black dress is 0·8.
The probability that Katharine will wear a black dress is 0·3.
a) Calculate the probability that both Helen and Katharine will wear black dresses.

b) Calculate the probability that at least one of them will wear a black dress.

13 A car dealer kept a record of cars he sold in September 2005.
One-fifth of the cars were yellow.
Two-sevenths of the cars were made in the UK.
Assume these events are independent.
If a car was chosen at random
a) what is the probability that it was yellow and made in the UK?
b) what is the probability that the car was either a yellow car made abroad or a non-yellow car made in the UK?

14 Sally has a bag containing cubes.
There are six white, seven red and three black cubes in the bag.
Sally takes out three cubes, without replacement.
What is the probability that all three are different colours?

KEY IDEAS

■ The probability of an event = $\dfrac{\text{number of ways the event can happen}}{\text{total number of possible outcomes}}$

■ P(an event happening) = 1 − P(the event not happening)

■ You use the addition rule to find the probability of either of two mutually exclusive events occurring. P(A or B) = P(A) + P(B)

■ You use the multiplication rule to find the probability of both of two events occurring. P(A and B) = P(A) × P(B)

■ If the outcome of an event is affected by the outcome of another event it is a dependent event.

■ You can use the multiplication rule for independent events and for dependent events, but you need to be careful about the values you use for the probabilities of dependent events.

Revision exercise D1

1 Sketch the graph of $y = \tan x$ for values of x from $^-90°$ to $450°$.

2 For what angles between $0°$ and $360°$ does $\tan x = 1$?

3 Sketch the graph of $y = \cos x$ for $0° \leqslant x \leqslant 360°$.
Given that one solution of $\cos x = ^-0.8$ is $143°$ to the nearest degree, find the other solution between $0°$ and $360°$.

4 Given that one solution of $\sin x = \frac{-1}{2}$ is $x = ^-30°$, use the symmetry of the graph of $y = \sin x$ to find all the solutions between $0°$ and $360°$.

5 Using a calculator and sketch graph, or otherwise, solve the equation $\cos x = 0.2$ for $0° \leqslant x \leqslant 360°$.

6 On the same set of axes, sketch the graphs of $y = \cos x$ and $y = \cos 2x$ for $0° \leqslant x \leqslant 360°$.

7 For $0° \leqslant x \leqslant 360°$, for what values of x does $\sin 2x = 1$?

8 This is the graph of $y = 2\sin 3x$.

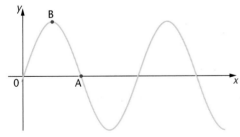

State the coordinates of A and B.

9 Sketch the graph of $y = 3\cos 2x$ for $0° \leqslant x \leqslant 360°$.

10 $f(x) = x^2 - 2$. Find the equation of the graph of $y = x^2 - 2$ after it has been translated by $\begin{pmatrix} 1 \\ 0 \end{pmatrix}$.

11 Describe the transformation which maps $y = g(x)$ on to each of these.
a) $y = g(3x)$
b) $y = 4g(x)$
c) $y = g(^-x)$

12 Sketch the graph of $y = \sin(4x)$ for values of x from $0°$ to $100°$.

13 a) Sketch the graph of $y = \sin(x + 90°)$.
b) State the equation of this graph more simply.

14 State the equation of the graph of $y = \cos x$ after
a) a translation of $\begin{pmatrix} 0 \\ 3 \end{pmatrix}$.
b) a one-way stretch parallel to the x-axis with scale factor 0.25.

15 The graph of $y = x^2$ is translated by $\begin{pmatrix} 3 \\ 1 \end{pmatrix}$ and then stretched with scale factor 2 parallel to the y-axis.
a) Find the equation of the resulting curve.
b) Find the coordinates of the minimum point on this curve.

16 A pack of cards contains five red, five blue, five green and five yellow cards. John takes two cards out without replacement.
What is the probability that
a) both cards are red?
b) the first card is red and the second card is blue?
c) at least one card is red?

17 On my way home from work I pass through three sets of traffic lights. The probabilities that I pass through them without stopping are 0.2, 0.4 and 0.7 respectively.
Find the probability that
a) I do not have to stop at any of the lights.
b) I have to stop at just one set of lights.
c) I have to stop at at least two sets of lights.

STAGE
10

Index

Index

STAGE
10